THE NATIONAL ACADEMIES SUMMIT ON

America's Energy Future

SUMMARY OF A MEETING

Committee for the National Academies Summit on America's Energy Future

NATIONAL RESEARCH COUNCIL
OF THE NATIONAL ACADEMIES

THE NATIONAL ACADEMIES PRESS
Washington, D.C.
www.nap.edu

THE NATIONAL ACADEMIES PRESS 500 Fifth Street, N.W. Washington, DC 20001

NOTICE: The project that is the subject of this report was approved by the Governing Board of the National Research Council, whose members are drawn from the councils of the National Academy of Sciences, the National Academy of Engineering, and the Institute of Medicine. The members of the committee responsible for the report were chosen for their special competences and with regard for appropriate balance.

Support for this project was provided by the Department of Energy under Grant Number DE-FG02-07-ER-15923 and by BP America, Dow Chemical Company Foundation, Fred Kavli and the Kavli Foundation, GE Energy, General Motors Corporation, Intel Corporation, and the W.M. Keck Foundation. Support was also provided by the Presidents' Circle Communications Initiative of the National Academies and by the National Academy of Sciences through the following endowed funds created to perpetually support the work of the National Research Council: Thomas Lincoln Casey Fund, Arthur L. Day Fund, W.K. Kellogg Foundation Fund, George and Cynthia Mitchell Endowment for Sustainability Science, and Frank Press Fund for Dissemination and Outreach. Any opinions, findings, conclusions, or recommendations expressed in this publication are those of the author(s) and do not necessarily reflect the views of the organizations that provided support for the project.

Library of Congress Cataloging-in-Publication Data

National Academies Summit on America's Energy Future (2008 : Washington, D.C.)
The National Academies Summit on America's Energy Future : summary of a meeting / Steve Olson and Robert W. Fri, editors.
p. cm.
"Committee for the National Academies Summit on America's Energy Future."
Chronicles the discussions of the summit held in Washington, D.C. March 13-14, 2008.
Includes bibliographical references.
ISBN 978-0-309-12478-2 (pbk.) — ISBN 978-0-309-12479-9 (pdf) 1. Power resources—United States—Congresses. 2. Energy policy—United States—Congresses. I. Olson, Steve, 1956- II. Fri, Robert W. III. Title.
TJ163.15.N365 2008
333.790973—dc22

2008044488

Available in limited supply from:

Board on Energy and Environmental Systems
National Research Council
500 Fifth Street, N.W.
Keck W917
Washington, DC 20001
202-334-3344

Additional copies of this report are available from the National Academies Press, 500 Fifth Street, N.W., Lockbox 285, Washington, DC 20055; (800) 624-6242 or (202) 334-3313 (in the Washington metropolitan area); Internet, http://www.nap.edu.

Printed on recycled stock

Printed in the United States of America

THE NATIONAL ACADEMIES

Advisers to the Nation on Science, Engineering, and Medicine

The **National Academy of Sciences** is a private, nonprofit, self-perpetuating society of distinguished scholars engaged in scientific and engineering research, dedicated to the furtherance of science and technology and to their use for the general welfare. Upon the authority of the charter granted to it by the Congress in 1863, the Academy has a mandate that requires it to advise the federal government on scientific and technical matters. Dr. Ralph J. Cicerone is president of the National Academy of Sciences.

The **National Academy of Engineering** was established in 1964, under the charter of the National Academy of Sciences, as a parallel organization of outstanding engineers. It is autonomous in its administration and in the selection of its members, sharing with the National Academy of Sciences the responsibility for advising the federal government. The National Academy of Engineering also sponsors engineering programs aimed at meeting national needs, encourages education and research, and recognizes the superior achievements of engineers. Dr. Charles M. Vest is president of the National Academy of Engineering.

The **Institute of Medicine** was established in 1970 by the National Academy of Sciences to secure the services of eminent members of appropriate professions in the examination of policy matters pertaining to the health of the public. The Institute acts under the responsibility given to the National Academy of Sciences by its congressional charter to be an adviser to the federal government and, upon its own initiative, to identify issues of medical care, research, and education. Dr. Harvey V. Fineberg is president of the Institute of Medicine.

The **National Research Council** was organized by the National Academy of Sciences in 1916 to associate the broad community of science and technology with the Academy's purposes of furthering knowledge and advising the federal government. Functioning in accordance with general policies determined by the Academy, the Council has become the principal operating agency of both the National Academy of Sciences and the National Academy of Engineering in providing services to the government, the public, and the scientific and engineering communities. The Council is administered jointly by both Academies and the Institute of Medicine. Dr. Ralph J. Cicerone and Dr. Charles M. Vest are chair and vice chair, respectively, of the National Research Council.

www.national-academies.org

COMMITTEE FOR THE NATIONAL ACADEMIES SUMMIT ON AMERICA'S ENERGY FUTURE

ROBERT W. FRI, Visiting Scholar, Resources for the Future, *Chair*
RICHARD A. MESERVE (NAE), President, Carnegie Institution for Science
LAWRENCE T. PAPAY (NAE), CEO and Principal, PQR, LLC
MAXINE L. SAVITZ (NAE), General Manager, Technology/Partnerships, Honeywell, Inc. (retired)
HAROLD T. SHAPIRO (IOM), President Emeritus and Professor of Economics and Public Affairs, Princeton University, and Chair, Committee on America's Energy Future, Technology Opportunities, Risks, and Tradeoffs

America's Energy Future Summit Staff

PETER D. BLAIR, Executive Director, Division on Engineering and Physical Sciences
JAMES ZUCCHETTO, Director, Board on Energy and Environmental Systems
MARTY A. PERREAULT, Director, Roundtable on Science and Technology for Sustainability
GREG EYRING, Senior Program Officer, Division on Engineering and Physical Sciences
DOROTHY MILLER, Christine Mirzayan Science and Technology Policy Fellow
LaNITA JONES, Program Associate, Board on Energy and Environmental Systems
STEVE OLSON, Consultant Writer
DEREK VOLLMER, Associate Program Officer, Science and Technology for Sustainability Program

America's Energy Future Project Staff

PETER D. BLAIR, Executive Director, Division on Engineering and Physical Sciences
JAMES ZUCCHETTO, Director, Board on Energy and Environmental Systems
KEVIN CROWLEY, Director, Nuclear and Radiation Studies Board; Staff Officer, America's Energy Future (AEF) Study Committee
JOHN HOLMES, Senior Program Officer, Board on Energy and Environmental System; Staff Officer, AEF Electricity from Renewables Panel

EVONNE TANG, Senior Program Officer, Board on Agriculture and Natural Resources; Staff Officer, AEF Alternative Liquid Transportation Fuels Panel
MADELINE WOODRUFF, Senior Program Officer, Board on Energy and Environmental Systems; Staff Officer, AEF Energy Efficiency Panel
GREG EYRING, Senior Program Officer, Division on Engineering and Physical Sciences, AEF Committee Coal, Oil, and Natural Gas Subgroup
SARAH CASE, Associate Program Officer, Board on Energy and Environmental Systems; AEF Committee Transmission & Distribution and Nuclear Subgroups
DOROTHY MILLER, Christine Mirzayan Science and Technology Policy Fellow
LaNITA JONES, Program Associate, Board on Energy and Environmental Systems, AEF Study Committee
JONATHAN YANGER, Senior Program Assistant, Board on Energy and Environmental Systems, AEF Panels

National Academies Office of Communications

BARBARA KLINE POPE, Executive Director, Communications and the National Academies Press
STEPHEN MAUTNER, Executive Editor, National Academies Press
ANN MERCHANT, Director, Outreach and Marketing
TERRELL SMITH, Senior Communications Officer

Foreword

A confluence of events is producing a growing sense of urgency about the role of energy in long-term U.S. economic vitality, national security, and climate change. Energy prices have been rising and are extremely volatile. The demand for energy has been increasing, especially in developing countries. Energy supplies, and especially supplies of oil, lack long-term security in the face of political instability and resource limits. Concerns about carbon dioxide emissions from the burning of fossil fuels, which currently supply most of the world's energy, are growing. Investments in the infrastructure and technologies needed to develop alternate energy sources are inadequate. And societal concerns surround the large-scale deployment of some alternate energy sources such as nuclear power. All of these factors are affected to a great degree by government policies both here and abroad.

To stimulate and inform a constructive national debate on these and other energy-related issues, the National Academy of Sciences and the National Academy of Engineering initiated in 2007 a major study, "America's Energy Future: Technology Opportunities, Risks, and Tradeoffs." The America's Energy Future (AEF) project was organized to respond to requests from the U.S. Congress, in particular from Senate Energy and Natural Resources Committee Chair Jeff Bingaman and Ranking Member Pete Domenici as well as House Science and Technology Committee Chair Bart Gordon and Ranking Member Ralph Hall. Phase I of the project is structured to provide authoritative estimates of the current contributions and future potential of existing and new energy supply and demand technologies, their associated impacts, and projected costs. It will also serve as the foundation for a Phase II portfolio of subsequent studies at the

Academies and elsewhere focused on more strategic, tactical, and policy issues, such as energy research and development priorities, strategic energy technology development, and policy analysis.

Phase I of the AEF project will produce a series of five reports designed to inform key energy policy decisions as a new U.S. President assumes office and a new Congress convenes in 2009. The AEF effort to date has benefited from a large number of recent projects conducted by various organizations that have explored technology options for shaping future energy use. Some of these study results conflict and reflect disagreements about technology potential, particularly for technologies such as biomass energy, energy efficiency, renewable electric power generating technologies, nuclear power, and advanced coal technologies. A key objective of the AEF series of reports is to resolve conflicting analyses of technology options to help facilitate a productive national policy debate about the nation's energy future.

The AEF project is being generously supported by the W.M. Keck Foundation, Fred Kavli and the Kavli Foundation, Intel Corporation, Dow Chemical Company Foundation, General Motors Corporation, GE Energy, BP America, the U.S. Department of Energy, and by our own academies.

A key milestone in the AEF project was the National Academies Summit on America's Energy Future, which was convened on March 13-14, 2008, in the National Academy of Sciences Auditorium in Washington, D.C. The summit provided an opportunity for discussion of recent major studies by key principals of those studies as input to the AEF study committee and panel deliberations. This summary report, the preparation of which was overseen by a subgroup of the Committee on America's Energy Future (see Appendix A), chronicles the rich and varied presentation that occurred at the summit. Information on the speakers at the summit is given in Appendix B, and the agenda for the summit is included as Appendix C.

Ralph J. Cicerone, President
National Academy of Sciences
Chair, National Research Council

Charles M. Vest, President
National Academy of Engineering
Vice Chair, National Research Council

Preface

On March 12, 2008, the price of a barrel of light crude oil exceeded $110 for the first time in history. The next day, more than 800 people gathered in the auditorium of the National Academy of Sciences Building and over the Internet for the 2-day National Academies Summit on America's Energy Future. While the summit was designed to examine a broad range of energy sources and timeframes ranging years and decades into the future, record-high prices of oil were a constant reminder that the future is fast approaching. As Department of Energy (DOE) Secretary Samuel Bodman said in addressing the summit, "The price of oil is so high that it has gotten everybody's attention."

The summit was conducted as the inaugural event in a major initiative at the National Academies, the America's Energy Future (AEF) project. A joint effort of several divisions within the National Academies, the AEF project has two phases. Phase I, an examination of the performance, costs, and potential impacts of existing and near-term energy technologies, will provide a base of information for a Phase II consideration of related policy issues such as managing climate change, ensuring energy (and, particularly, oil) security, and developing and deploying advanced technologies that will help meet those challenges. The Phase I effort will culminate in a full study report by the Committee on America's Energy Future, supported by three separately appointed panels whose work will be detailed in three separate reports (see Appendix A).

Conceived and carried out as a collaborative effort between the AEF project and the National Academies Office of Communications, the summit was held to stimulate discussion and debate in advance of the 2008 U.S. elections.

It brought together many of the most knowledgeable and influential people working on energy issues today (see Appendix B). In addition to Secretary Bodman, government speakers included Senator Jeff Bingaman, DOE Undersecretary Ray Orbach, and Department of State Undersecretary Reuben Jeffery. James Schlesinger, the first secretary of energy, brought a valuable historical perspective to the meeting, while Ged Davis from the World Economic Forum in Geneva and José Goldemberg of Brazil offered international perspectives. Several speakers who have recently led important national and international studies on energy issues—including Robert Marlay (Climate Change Technology Program), Rodney Nelson (National Petroleum Council), Ernest Moniz (Massachusetts Institute of Technology), Paul Portney (University of Arizona), Michael Ramage (ExxonMobil [retired]), John Holdren (Harvard University), Steven Chu (Lawrence Berkeley National Laboratory), and Amory Lovins (Rocky Mountain Institute)—summarized and elaborated on their previous work. Other representatives of higher education, industry, and the nonprofit sector provided informative and provocative analyses of a very broad range of issues.

The result was an incredibly rich gathering of intellectual capital. Full biographies of the speakers, videos of their presentations, and copies of their slides are available at the AEF project website, http://www.nationalacademies.org/energy. The website also contains links to other energy-related activities and to many reports and other documents available from the National Academies.

The present report, the first in the AEF Phase I series of five reports, was prepared by Steve Olson in close collaboration with a subset of the Committee on America's Energy Future. It summarizes what was discussed at the summit but cannot cover all topics that the committee thinks are important. Nor does the report necessarily reflect the views of the committee. The report is organized thematically rather than chronologically, and so points made by some speakers appear in more than one chapter. It also reproduces some of the slides shown at the summit, some of which have been slightly altered for clarity.

Even over the course of 2 days, not all of the topics associated with energy could be discussed at the summit. For example, renewable sources of energy were covered briefly but not thoroughly, and international considerations received less attention than did U.S. policies and scenarios—gaps in coverage that are necessarily reflected in this summary. Nevertheless, the speakers at the summit covered a very broad range of topics, from economic development in China to ethanol production in Brazil to the anticipated effects of recent U.S. legislation on greenhouse gas emissions, in talks loosely organized around three major themes: energy security, energy and the economy, and energy and the environment (see Appendix C). In particular, several speakers examined recent actions and the need for future actions in the context of calls for policy reforms from major national organizations. They pointed out that some progress has been made on some issues. Yet the challenges facing the United States, other developed nations, and the developing world remain immense. Meeting the need for energy without irreparably damaging Earth's environment will require technological and social changes that have few parallels in human history.

As this summary makes clear, the energy problem is hard—much harder than projects with straightforward technological objectives like the Manhattan Project or the Apollo Project. As Senator Jeff Bingaman said at the summit, "Energy policy does not have a single goal. It is extremely complex and multifaceted. . . . We run a real risk of heading in the wrong direction in energy policy if we try to oversimplify the issues, if we try to overstate the potential of any single energy initiative, or if we try to understate the difficult nature of the energy problems that we face."

Despite such difficulties, the discussions at the summit were largely optimistic. Many advanced technologies outlined at the summit hold great promise. There are immense markets for green technologies around the world. International cooperation can help to unify the efforts of national governments. The challenge is to convert concern into action. The energy problem is solvable, but not without tremendous effort and good will on the part of individuals, organizations, and nations.

Robert W. Fri, Chair
Committee for the National Academies Summit on America's Energy Future

Peter D. Blair, Executive Director
Division on Engineering and Physical Sciences, National Research Council

Acknowledgments

This report has been reviewed in draft form by individuals chosen for their diverse perspectives and technical expertise, in accordance with procedures approved by the National Research Council's Report Review Committee. The purpose of this independent review is to provide candid and critical comments that will assist the institution in making its published report as sound as possible and to ensure that the report meets institutional standards for objectivity, evidence, and responsiveness to the study charge. The review comments and draft manuscript remain confidential to protect the integrity of the deliberative process. We wish to thank the following individuals for their review of this report:

R. Stephen Berry (NAS), University of Chicago,
Jan Beyea, Consulting in the Public Interest,
Marilyn A. Brown, Georgia Institute of Technology,
Douglas M. Chapin (NAE), MPR Associates, Inc.,
Ged Davis, International Institute of Applied Systems Analysis,
Mildred R. Dresselhaus (NAS/NAE), Massachusetts Institute of Technology,
William Fulkerson, University of Tennessee, and
Kurt Yeager, Electric Power Research Institute.

Although the reviewers listed above have provided many constructive comments and suggestions, they were not asked to endorse any statements made by speakers at the summit, nor did they see the final draft of the report before its release. The review of this report was overseen by Elisabeth M. Drake (NAE), Massachusetts Institute of Technology, and Robert A. Frosch (NAE), Harvard University. Appointed by the National Research Council, they were responsible for making certain that an independent examination of this report was carried out in accordance with institutional procedures and that all review comments were carefully considered. Responsibility for the final content of this report rests entirely with the committee and the institution.

The National Academies Summit on America's Energy Future was a major event involving many individuals across the National Academies organization whose contributions are gratefully acknowledged. These individuals include Clinton Alsip, John Bavier, Chris Benson, Katherine Bittner, Sheryl Bottner, Virginia Bryant, Dana Caines, Mark L. Carter, Sumana Chatterjee, Annie Drinkard, Molly Galvin, Penelope Gibbs, Sylvia Gilbert, Sally Groom, John Horan, Julie Ische, Jim Jensen, Maria Jones, William Kearney, Patrice Legro, Dorothy Lewis, Matthew Litts, Alphonse MacDonald, Rachel Marcus, Scott Maslin, Francesca Moghari, Shellie Myers, Neo Pardo, Patsy Powell, Dustin Pusch, Moises Ramirez, Cortney Riese, Barbara Schlein, Olive Schwarzschild, Sharon Segal, Brett Simmons, Priya Sreedharan, Ariel Suarez, Mario Velasquez, Chris Watson, Sue Wyatt, and Johann Yurgen.

Contents

PART II
ENERGY SUPPLIES

PART III
ENERGY USES

PART IV
MEETING THE CHALLENGE

APPENDIXES

Part I

The Current Context

We must find a way to meet the increasing demand for energy without adding catastrophically to greenhouse gases.

—Ray Orbach, Undersecretary for Science,
U.S. Department of Energy

1

A Growing Sense of Urgency

In 2003 the National Academy of Engineering named the 20 most important engineering accomplishments of the 20th century. At the top of the list was electrification. As Ernest Moniz said at the Summit on America's Energy Future, which was held March 13-14, 2008, at the National Academy of Sciences Building in Washington, D.C., "In the century of computers, lasers, rockets, and cell phones, for electrification to be the number one engineering achievement of the century reflects the enormous technology embedded in that system, and its implications for our quality of life and the building of our society."

The use of energy permeates our lives. We use energy to cook, to light and heat our homes and commercial buildings, to power industry and agriculture, to transport people and goods, and to drive and fly. The products we buy and the services we employ are made possible by the use of energy. Our well-being, prosperity, and security are all built on the ingenious provision and application of various forms of energy.

These observations apply to those who have relatively limited access to energy as much as they do to those who have ready access. As Reuben Jeffrey noted during the summit, the growing use of energy in the developing world is closely linked to economic development and the reduction of poverty. Economic growth in China, for example, has lifted hundreds of millions of people out of poverty and has brought hope and opportunity to a new generation. The United States has an interest in supporting this economic development, Jeffrey said, while also leading the effort to find ways of mitigating the negative environmental impacts that accompany development.

Steven Chu made a similar observation. The major energy problem in about one-third of the world is very different from the energy problems common in the United States, he said: "The basic energy needs of the poorest people are not being met." The poor in the developing world rely primarily on wood, charcoal, dung, or other organic materials for cooking, and most live without electricity (IAC, 2007). The rest of the world has a "moral and social obligation" to help those who live in poverty gain access to the energy they need, Chu said.

Yet continued reliance on the dominant sources of energy being used today also poses grave risks to human well-being. The production and use of fossil energy cause air and water pollution and also can require huge economic investments. Efforts to secure long-term access to fossil and nuclear fuels have at times greatly exacerbated international tensions. And global climate change caused by the release of greenhouse gases into the atmosphere could have catastrophic consequences. "There is a broad consensus that if a responsible society doesn't act, we're heading for a problem in the environment, on energy security, and on economic development," said Jeffrey. "We need to act now."

INCREASING DEMAND AND CONSTRAINED SUPPLIES

In his 12-volume *Study of History* (1934-1961), historian Arnold Toynbee examined the trajectories of various human civilizations and asked why civilizations succeeded or failed. Most often, said James Schlesinger at the summit, they failed because of a challenge they could not meet. Today, the provision and use of energy pose "an immense challenge, both foreign and domestic," Schlesinger said. "The question is our ability to respond effectively to that challenge."

Schlesinger quoted from a 1953 book entitled *The Next Million Years* by Charles Galton Darwin, the grandson of the author of the *Origin of Species*. In 1953 Darwin wrote:

> A thing that will assume enormous importance quite soon is the exhaustion of our fuel resources. Coal and oil have been accumulating in the earth for over five hundred million years, and, at the present rates of demand for mechanical power, the estimates are that oil will be all gone in about a century, and coal probably in a good deal less than five hundred years. For the present purpose, it does not matter if these are under-estimates; they could be doubled or trebled and still not affect the argument. Mechanical power comes from our reserves of energy, and we are squandering our energy capital quite recklessly. It will very soon be all gone, and in the long run we shall have to live from year to year on our earnings. (p. 63)

Darwin's observations are as relevant today as they were a half century ago, Schlesinger observed. "We are going through our capital of inheritance at a remarkable pace."

Several years ago Schlesinger co-chaired a Council on Foreign Relations study with John Deutch of the Massachusetts Institute of Technology on the geopolitical implications of America's dependency on imported oil (Victor et al., 2006). That study articulated three key observations, which several other speakers at the summit reiterated. The first is that the rising consumption of oil (driven by rising incomes and population growth) in the face of limited supplies will continue to put upward pressure on prices. This development is a departure from the past, said Schlesinger, when bodies such as the Organization of Petroleum Exporting Countries were established to limit the abundance of supplies.

Current demand for energy services reflects both the large amount of energy being used in the industrialized countries and the increasing use of energy in the developing world (Chapter 3). Consumption of petroleum, natural gas, and coal in the United States, Europe, and industrialized Asian countries is large but relatively stable, Jeffrey pointed out. But in rapidly developing countries such as China and India, energy consumption is growing dramatically. As recently as the early 1990s, China supplied its own energy resources. In less than two decades, China's oil consumption and its gross domestic product have tripled. As a result, China now imports more than 3 million barrels of oil per day, and that number is growing rapidly.

Demand for energy is likely to increase even more in the years ahead (Figure 1.1). The world's energy consumption could grow by more than 50 percent by 2030, with three-quarters of that increase coming from the developing world. By the end of the 21st century, energy use could more than double, although actual use will depend on prices, availability, new technologies, and many other factors.

Meeting a rising demand for energy will require significant investment, Jeffrey pointed out. The International Energy Agency (IEA) estimates that $22 trillion in new investment will be needed to meet expected global demand in the next two decades (IEA, 2007).

Production and the discovery of new energy sources are not yielding sufficient fuels and electricity to keep pace with increasing demand. With regard to petroleum (Chapter 4), reserves of oil are available, but many of the countries with petroleum reserves do not have a strong incentive to increase production. "We're not dealing with an absence of oil," said Samuel Bodman. "We're dealing with the fact that a large fraction of the oil that is in the world—and we know where it is—is in the hands of national oil companies that are seeing $110 a barrel today for the price of oil, and they are looking at that price going up. They are starting to ask the question, in my judgment, why should they produce it now when they can produce it next year and make more money?"

Supplies of natural gas also are constrained. Russia, which has the world's largest known natural gas reserves, has indicated that it will not develop the Yamal gas field fast enough to meet the growing demand in Western Europe,

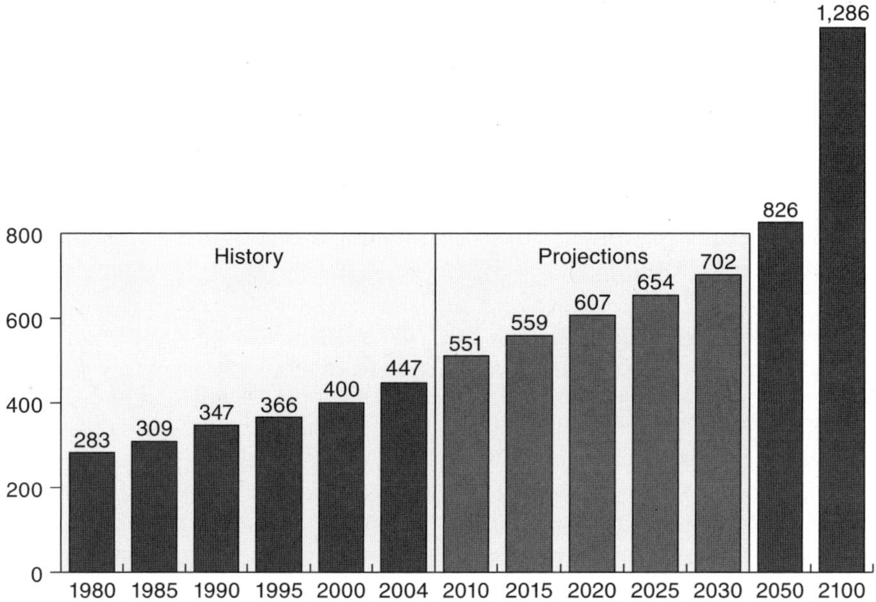

FIGURE 1.1 World energy consumption (expressed in terms of quadrillion British thermal units; quads) is projected to increase by more than 50 percent between now and the year 2030, and by more than 100 percent by the end of the 21st century. NOTE: Projections are based on models that incorporate assumptions about such factors as economic growth, world energy markets, resource availability and costs, technologies, and demographics. A review of the Energy Information Administration's process for energy forecasting is available at http://www.eia.doe.gov/oiaf/aeo/overview/index.html. SOURCE: Ray L. Orbach, U.S. Department of Energy, "Basic Research and America's Energy Future," presentation at the Summit on America's Energy Future on March 14, 2008; data on consumption through 2030 from EIA (2007) and after 2030 through 2100 from IPCC (2000; Figure 8).

Schlesinger said. Iran, which has the second largest known reserves, also cannot be expected to expand production greatly.

Supplies of liquefied natural gas are not a problem in the short term, according to Schlesinger. But in the long run, depending on these supplies implies reliance on the same countries that supply much of the world's oil. Liquefied natural gas also will not be cheap. European contracts for liquefied natural gas track the price of diesel, Schlesinger said, and with diesel prices high the price of liquefied natural gas also is high.

Coal (Chapter 5)—once considered America's "ace in the hole" because of abundant domestic supplies—now poses much greater uncertainty, Schlesinger observed, largely because of concern about increasing levels of greenhouse gases in the atmosphere (Chapter 2). Although the Energy Information Administration (EIA) had projected that coal would provide 57 percent of U.S. electrical power production by 2030 (EIA, 2008)—up from 51 percent today—recent cancellations of new coal-fired power plants make such projections unlikely. Technologies are being investigated that would capture the carbon dioxide emitted from coal-fired plants and sequester it in deep underground reservoirs (Chapter 5). But any such technology is years away from being deployed on a large scale.

Nuclear power (Chapter 6) is receiving renewed attention as a potential source of energy that does not release greenhouse gases during power generation. But while a nuclear revival is likely, Schlesinger said, a nuclear renaissance is not. Major problems regarding fuel cycles and waste storage remain to be solved. And the nuclear labor force has aged dramatically in recent decades, with much of the labor force now on the verge of retirement and far fewer trained workers ready to fill their shoes. The EIA (2008) has projected that by 2030 nuclear energy will provide an additional 14.7 gigawatts of power in the United States (current nuclear plants have a capacity of about 100 gigawatts). That will be helpful, Schlesinger said, but it will not solve America's energy problem.

Energy supplies will continue to be available, said Schlesinger, but only at whatever price clears the market. As a result, "the price of energy will continue to rise," he said, "[despite] what the public expects and what political leaders may promise with regard to affordability."

CONTINUED U.S. RELIANCE ON FOREIGN SOURCES OF OIL

The second broad observation to emerge from the study Schlesinger co-chaired is that U.S. dependence on foreign oil is not going to end in the foreseeable future (Victor et al., 2006). One reason for this continuing reliance is the sheer size of the U.S. market. As Paul Portney pointed out, about one person in every 20 on Earth lives in the United States. Yet the U.S. population currently uses about one in every four barrels of oil that are produced. This high usage is why the United States currently imports about two-thirds of its oil. "We use oil in gross disproportion to our numbers," Portney said.

Past calls to achieve energy independence have failed, Schlesinger observed. In 1973, President Richard Nixon launched Project Independence, which laid out a path to becoming energy self-sufficient by the year 1980. Between 1973 and 1980, imports of crude oil into the United States rose 60 percent, Schlesinger said. Since that time, they have tripled. "If we are seeking energy independence, we do not seem to be on the right track," Schlesinger observed.

The main reason for increased imports is the nation's continued reliance on petroleum for vehicle fuels. "We are not going to reach energy independence as long as the United States remains dependent on the internal combustion engine," said Schlesinger. Advances in transportation technologies have substantial potential to reduce oil consumption in this sector (Chapter 9). But even as the fuel efficiency of vehicles improves, the larger number of vehicles caused by a larger population and more vehicles per person will at least partly offset efficiency gains. Furthermore, it takes 20 years to turn over the stock of cars, and older cars, which tend to stay on the road in the absence of policies to replace them, are the least efficient.

Because the United States will not achieve energy security in the foreseeable future, Schlesinger said, it must instead strive to fashion a set of policies that will limit energy insecurity. Establishing the strategic petroleum reserve was an important step in that direction, he commented, but much more needs to be done.

GROWING INTERNATIONAL CONSEQUENCES

The third observation to emerge from the Council on Foreign Relations study is that the continued U.S. reliance on imported oil has immense international consequences (Victor et al., 2006). Portney underscored this point when he remarked that, with oil at more than $100 per barrel, the United States is sending about $500 billion per year to oil-producing countries in return for needed supplies of oil. This outflow of funds contributes to an enormous trade deficit, which in turn is putting downward pressure on the value of the dollar. As a result, imports have become more expensive. And because the United States has not succeeded in greatly increasing its own exports to other countries and has chosen to run a large budget deficit, the nation has to rely on other countries buying U.S. bonds and other assets to finance its trade deficit.

The bottom line is that debt payments to other countries are growing and will continue to do so. "For my children and your children and their children, if we continue down this road, we're going to see a future in which large amounts of the government budget each year go to servicing the debt by making payments to people in China or the Middle East or Venezuela or Nigeria," said Portney. "That's not becoming of a country of the stature of the United States of America in my view."

The U.S. dependence on other countries for oil has weakened the nation's ability to influence the policies of those countries, said Schlesinger. The growth of oil revenues in countries like Iran and Venezuela has enabled them to be openly defiant with the United States in ways that would not have occurred a decade ago. Russia is reasserting itself after a decade of what it views as international humiliation. Even Saudi Arabia and other U.S. allies in the Gulf of Arabia are not as responsive to U.S. requests as they once were because of the immense increase in their financial assets.

Also, many of the countries on which the United States relies for petroleum are located in unstable parts of the world, Portney observed. In addition, the rise of resource nationalism and the use of energy resources as a political tool have put constraints on oil production. More than 75 percent of the world's oil reserves are now controlled by national oil companies, which tend to be less efficient at developing their resources (NPC, 2007).

Transporting supplies to market also can be a problem, Jeffrey pointed out. Much of the world's oil passes through a series of vulnerable points, leaving energy consumers susceptible to supply disruptions. Thirty-five percent of all the oil shipped in the world passes through the Strait of Hormuz. Another 34 percent passes through the Strait of Malacca on the way to China, Japan, and other Asian countries. About 8 percent passes through the Bab el Mandeb, which connects the Red Sea to the Gulf of Aden. "Reliance on a small number of oil transit routes leaves world oil supplies open to the possibility of a sudden disruption, whether manmade or by a natural disaster," Jeffrey said.

Oil facilities also are attractive targets for terrorists, Schlesinger said. Al Qaeda members who were planning attacks on oil facilities in the Middle East were recently arrested in Saudi Arabia—in February 2007 Saudi forces thwarted an attack on the Abqaia oil processing center, which handles two-thirds of the oil supply from Saudi Arabia. "And," said Schlesinger, "Al Qaeda has not given up."

High oil prices have a significant impact on all energy-consuming countries. According to Jeffrey, they cause hardship for individuals who have to pay higher prices for energy, and they also have a macroeconomic effect. Many other countries besides the United States are net oil importers, including some of the world's poorest and most vulnerable countries. In many of those countries, the outflow of money to pay for oil risks diverting resources from public services and crucial tasks such as economic development. And major economic disruptions of recent decades, such as the developing countries' debt crisis of the 1980s, were exacerbated by financial imbalances related to energy imports and exports.

Rising energy prices have a particularly damaging impact on small, developing countries, Bodman pointed out. There, high energy prices can be especially pernicious in stifling business growth and slowing improvements in the quality of life. "We must keep the energy needs of the world's poorest nations in mind as we talk about solutions," he said.

On the other side of the ledger, oil-exporting countries are accumulating huge financial surpluses. One consequence of this wealth transfer is a proliferation of sovereign wealth funds, which are large pools of money that countries rather than private entities invest. The United States and other countries are working with the International Monetary Fund and the Organisation for Economic Co-operation and Development to identify best practices to encourage greater transparency in the investment of these funds, which will help ensure that such investments are based on market principles. The current administration supports policies that keep the United States open to investments from abroad, Jeffrey said, while ensuring that national security is not threatened.

A NECESSARY URGENCY?

Despite the severe problems associated with energy production and use, many people at the summit expressed optimism that the problems can be overcome. Many new technologies are already available that can reduce energy consumption in transportation (Chapter 9) and in buildings and industry (Chapter 10), and other promising technologies are being developed. Moreover, many more people are recognizing the urgency of the energy issue, Bodman observed, which has built support for one of the most important elements of a national strategy: a national imperative to act. "Perhaps as never before, the American people are calling for action," he said.

Yet many speakers at the summit also asked whether the level of urgency being expressed by the public and by policymakers is sufficient. New technologies can take a long time to develop and implement, especially given the large investments that must be made for new technologies to have a substantial effect in the energy sector. In light of projections that call for the use of energy to double during the 21st century, said Ray Orbach, "I think one can legitimately ask, 'Where will the energy come from?'" Simple extrapolations of current trends into the future appear to yield untenable increases in greenhouse gas concentrations in the atmosphere. "Can we survive as a globe under those conditions?" Orbach asked.

"We don't seem to be able to generate the sense of urgency that's required to address this problem," said Ernest Moniz. "We talk about doing this and doing that, and before you know it a decade has passed. We can't afford to waste another 10 or 15 years. We can't get there from here if we do that. The sense of urgency is one that we need somehow to capture."

2

The Challenge of Global Warming

According to the available evidence, the second half of the 20th century was warmer than any other 50-year period in the last 500 years, and probably in the last 1,300 years, Ged Davis said at the summit. The 20th century saw about a 0.6-degree centigrade increase (about 1.1 degrees Fahrenheit) in global and ocean temperatures (Figure 2.1). Over that same period, sea level rose about 150 millimeters (6 inches), and it is continuing to rise at about 3 millimeters (an eighth of an inch) per year. Mountain glaciers and snow cover have on average declined in both hemispheres.

These global changes appear to be a consequence of changes in land use and energy use during the 20th century, Davis said. Deforestation and the burning of fossil fuels have increased the amount of carbon dioxide in the atmosphere from about 300 parts per million in 1900 to about 380 parts per million today. This increase has been driven by a fourfold increase in global population—to more than 6 billion—combined with an increase in the per capita use of energy. Given that per capita global incomes have risen some 10-fold since the beginning of the 20th century, the world has seen a 40-fold increase in economic activity. Energy use has not risen as fast as economic activity, owing to increases in efficiency and changes in the nature of economic activity. But Davis estimated that annual global energy use has probably risen at least 20-fold over the past century.

The Intergovernmental Panel on Climate Change (IPCC) models that predict temperatures based only on natural forcings—such as changes in solar output and volcanic activity—show relatively moderate temperature changes over the 20th century (Figure 2.2). IPCC models that include the effects of

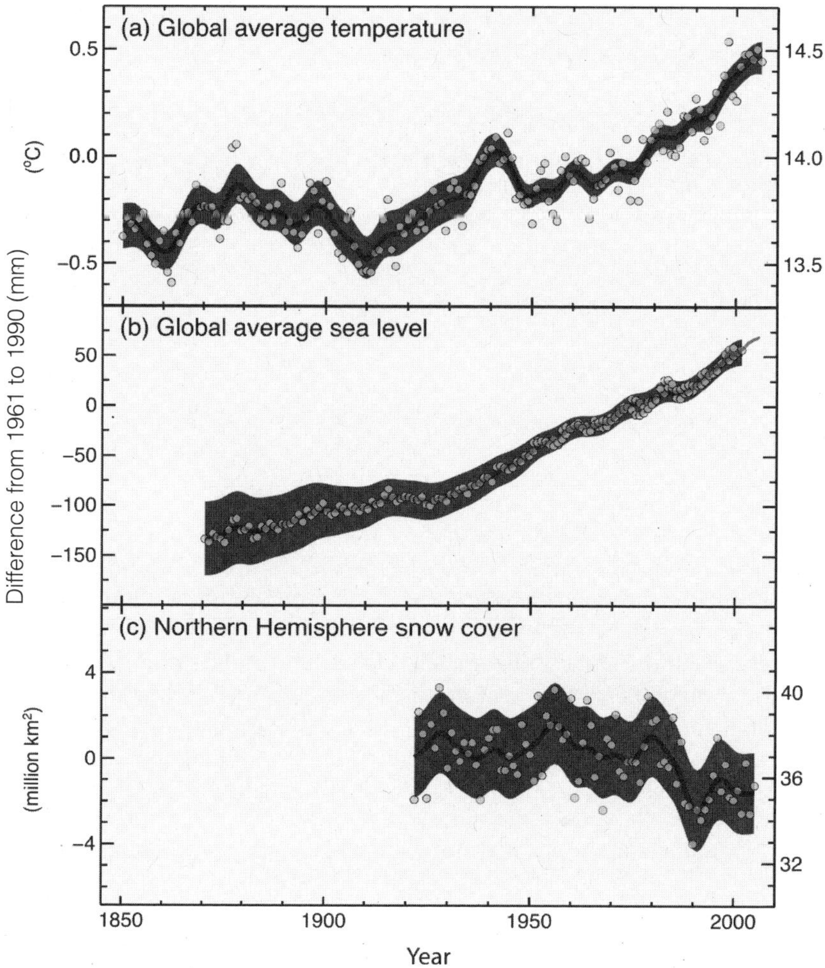

FIGURE 2.1 Global air and ocean temperatures and sea level have risen over the course of the 20th century, while average annual snow cover in the Northern Hemisphere has declined. SOURCE: IPCC (2007; Figure 1-1).

anthropogenic greenhouse gas emissions show an increase in temperatures that closely tracks observations.

SCENARIOS OF FUTURE CLIMATE CHANGE

On the basis of models that assume different technology, economic, and policy trajectories, the IPCC has developed several scenarios of future emis-

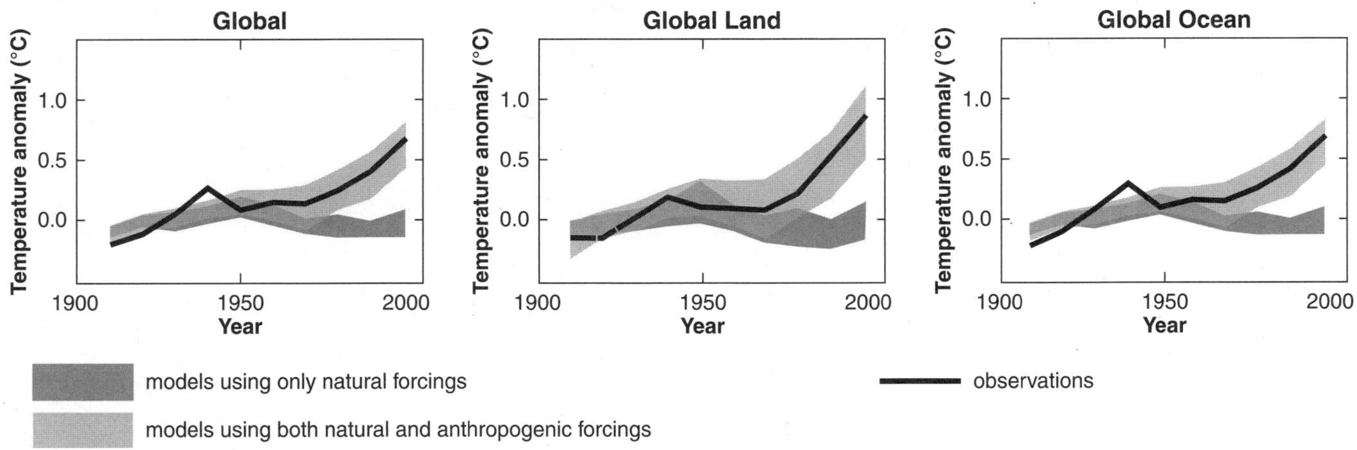

FIGURE 2.2 Models that incorporate both natural and human causes of climate change more accurately predict temperatures over the past century as shown by the fit between the predictions of such models (upper bands) and observed temperatures (black lines). SOURCE: IPCC (2007; Figure 2-5).

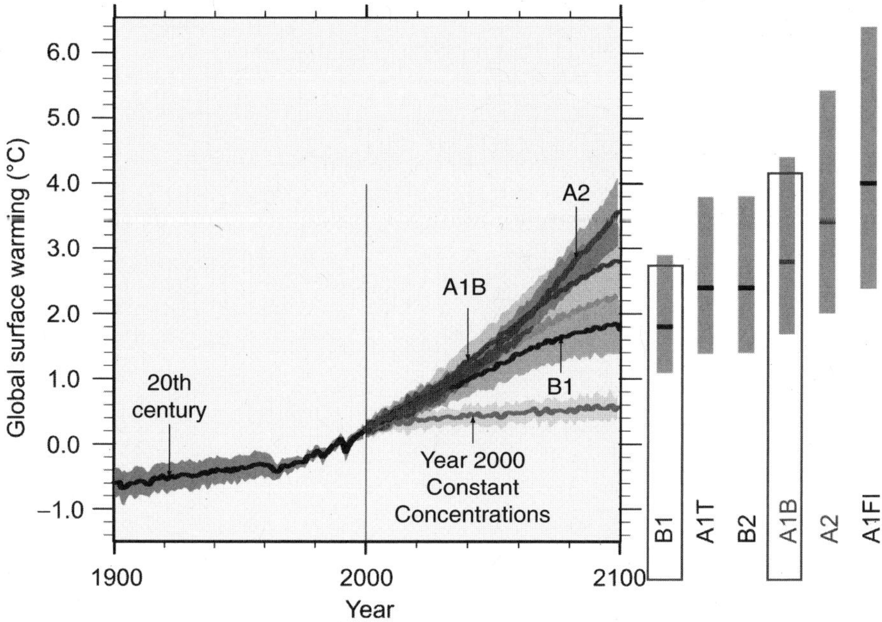

FIGURE 2.3 Temperature increases in the 21st century are very likely to be larger than those observed in the 20th century according to scenarios developed by the Intergovernmental Panel on Climate Change. For definitions of scenarios B1 and A1F1 at each end of the range, see the main text. SOURCE: IPCC (2007; Figure 3-2).

sions (Figure 2.3). A low-emission scenario, B1, assumes a mid-century peak in global population, the rapid development of a services-oriented economy, and a change toward clean and efficient energy technologies. A high-emission scenario, A1FI, assumes a mid-century peak in population, rapid economic growth, and intensive use of fossil fuels for energy. Other scenarios fall between those two extremes.

Even if all use of fossil fuels were to cease today, these models predict another 0.6-degree centigrade increase in temperature during the 21st century, Davis observed. Since all of the IPCC scenarios assume continued use of fossil fuels, all of the scenarios assume temperature increases larger than that amount.

The most positive scenario (B1) results in model predictions of a 1.5- to 2-degree centigrade (2.7- to 3.6-degree Fahrenheit) temperature increase over the 21st century. This scenario is almost certainly overoptimistic, according to Davis. With a more balanced mix of assumptions (scenario A1B), an additional 1-degree centigrade (1.8-degree Fahrenheit) increase occurs. As a result, the models predict a 2- to 3-degree centigrade (3.6- to 5.4-degree Fahrenheit) increase in global temperatures.

In the most extreme scenarios, the models predict that temperature increases would be much larger. If China and India rely heavily on coal until the middle of the century for electricity, temperatures could, the models predict, go up 5 degrees centigrade (9 degrees Fahrenheit) or more.

Temperature increases are, in turn, predicted to have major impacts on water, ecosystems, food, coastal areas, and health, Davis observed (Figure 2.4). Water supplies are predicted to dwindle in some areas, which could lead to human suffering and interregional tensions. Many ecosystems could be devastated, greatly reducing the benefits those ecosystems provide to human societies. Agriculture could suffer in areas where the climate becomes more severe, coastal areas could be more vulnerable to flooding and loss of wetlands, and the burden of disease could grow. Already, problems are intensifying in mountainous regions, the Arctic, regions subject to drought, and low-lying areas.

Davis emphasized that some uncertainties continue to surround the projections of temperature increases. "Models are evolving and developing, and we're dealing with complex dynamic systems that play out over one or two centuries," he said. These uncertainties could lead to less warming than expected, but they also could lead to more.

Also, other possible environmental effects linked to greenhouse gas emissions could prove to be serious, even if they are not much discussed currently. For example, Richard Meserve noted that about one-third of the carbon dioxide that is released into the atmosphere is absorbed into the oceans. When it is absorbed, it is converted into carbonic acid, which increases the acidity of the oceans. This acidification of the ocean interferes with the ability of some organisms to take up calcium carbonate and use it in their shells and skeletons. Over time, this effect could provoke a worldwide crisis for corals and other organisms. The acidification of the ocean is already being observed, and models predict that this process will continue throughout the 21st century.

Another potential unanticipated consequence involves the biological productivity of the oceans. Satellite readings of the oceans have revealed that large areas of the ocean have less plankton than they did previously, Meserve said. These "oceanic deserts" appear to result from reduced upwelling of deep nutrient-rich waters to the surface. Each year the area so affected increases by about the size of Texas. "The problem is growing more severe, with adverse impacts that are perhaps somewhat different and certainly occurring much more quickly than we had perhaps anticipated as recently as 3 or 4 years ago."

STEPS TO BE TAKEN

Without major policy changes, emissions of carbon dioxide will continue to increase in future years. In the IPCC scenario that assumes no policy interventions for reducing greenhouse gas emissions (the "reference scenario" in Figure 2.5), the total global emissions of carbon dioxide as a result of energy

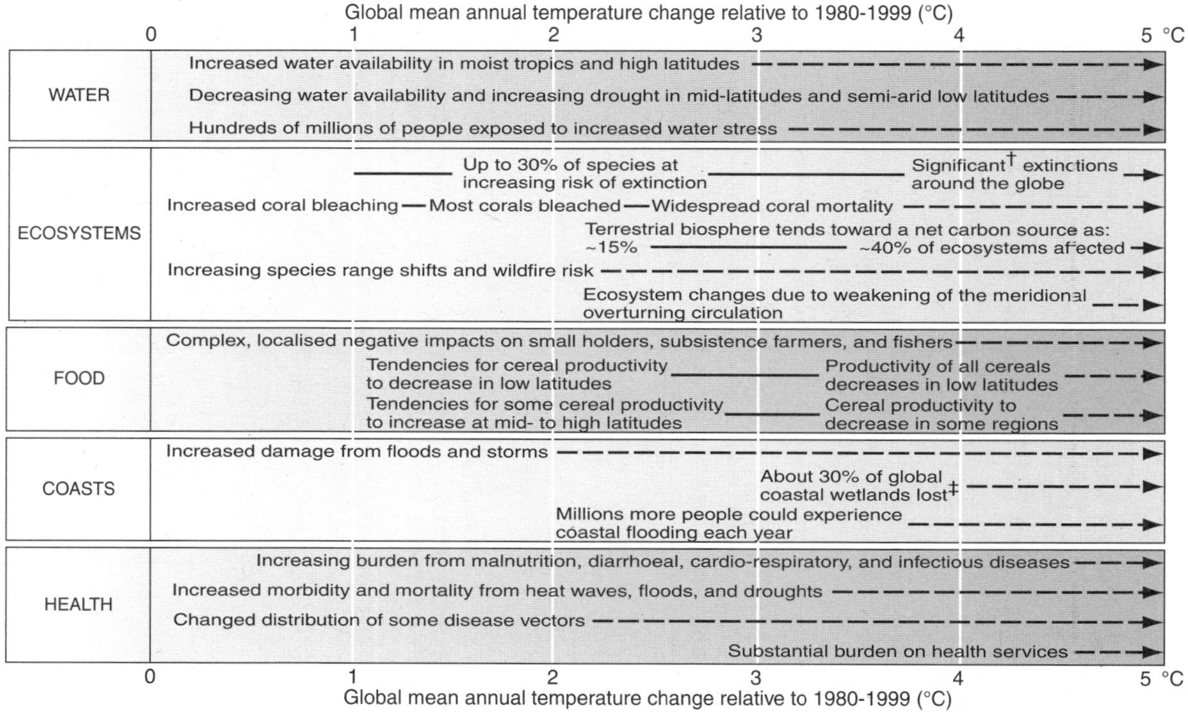

FIGURE 2.4 Temperature increases will have major impacts on the environment, agriculture, and human health. †"Significant" is defined here as more than 40 percent. ‡Based on average rate of sea level rise of 4.2 millimeters per year from 2000 to 2080. SOURCE: IPCC (2007; Figure 3-6).

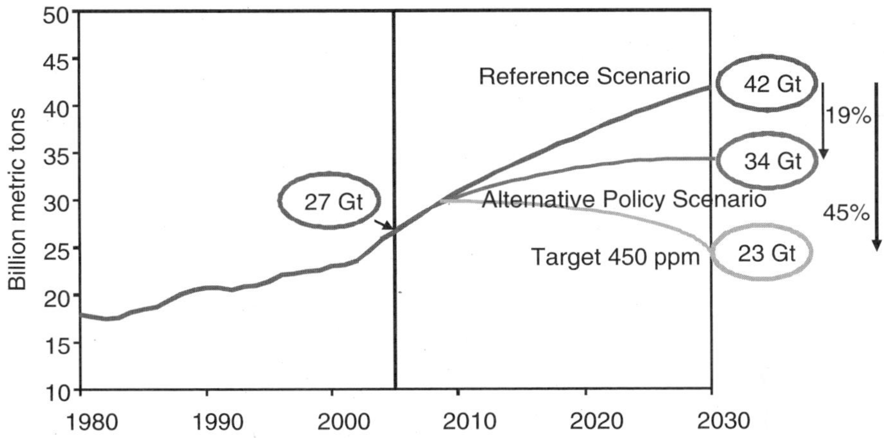

FIGURE 2.5 Without governmental policies to protect against climate change, carbon dioxide emissions will increase from 27 billion metric tons (shown as Gt) today to 42 billion metric tons by 2030. SOURCE: IEA (2007).

use rise from 27 billion metric tons today to 42 billion metric tons by 2030. Furthermore, the current mix of energy sources in most countries is unlikely to change, Davis observed. Indeed, in some countries, the percentage of fossil fuels in their energy mix is likely to increase.

Various policy initiatives discussed in the rest of this summary will be needed to reduce carbon dioxide emissions. Stabilizing concentrations of carbon dioxide in the atmosphere will be an even greater challenge, Davis said. Emissions reductions that occur as a result of the Energy Independence and Security Act of 2007 (which is discussed in part IV of this summary) are "just a little drop in the ocean."

To avoid dangerous levels of climate change, action must begin soon, Davis said, especially given the investments that must be made in energy infrastructure. Retrofitting equipment that is already in place will be very expensive. At least $1 trillion of investment will be needed per year to achieve a low-carbon future, according to estimates Davis cited. In a $60 trillion per year global economy, that investment may sound achievable. But "many did not expect to put that sort of money into energy capital," Davis pointed out.

Furthermore, trends in the release of greenhouse gases are headed in the wrong direction. Meserve described a recent study (Raupach et al., 2007) that found that the rate at which carbon dioxide is accumulating in the atmosphere accelerated from about 1 percent during the 1990s to an estimated 3 percent since the year 2000. According to the study, 60 percent of that increase is the result of increased global economic activity. Twenty percent is due to increased energy intensity in developing nations, which are using more fossil fuels and are using those fuels less efficiently than other countries. And the final 20 percent is accounted for by the troubling observation that sinks responsible for absorbing carbon dioxide from the atmosphere appear to be less effective today than they were in the past. For example, changes in wind patterns in the Southern Hemisphere have caused carbon-rich water to stay close to the surface, which means that less carbon dioxide can be absorbed into the oceans. Similarly, the land surface seems to be taking up less carbon dioxide than it has in the past. "The bottom line is that we are not making progress in reducing carbon dioxide emissions," said Meserve. "The problem is getting worse at an accelerating rate over time, rather than better."

THE RELEVANT TIME SCALES

Davis and several other speakers at the summit emphasized that issues associated with energy need to be viewed in the context of different time scales. For example, understanding climate change and its implications requires consideration of periods of 100 to 200 years. "Policymakers have never had to take that seriously into account in framing policy," Davis said.

The most important timeframe for policy choices, Davis said, is the period between the present and 25 to 35 years from now. During that time, major components of policy can change, and the dynamics of policy change over that period are somewhat understood. For example, the world will need to go through two major transitions before the middle of the century. Oil production is likely to enter a long plateau within a decade or two, Davis predicted. Conventional natural gas, about which somewhat less is known, similarly will move toward a plateau before 2050 and maybe earlier, said Davis. The world will have to traverse those two transitions while developing and implementing a new set of technologies for the second half of the century.

Over the next 25 to 35 years, the changing geopolitical context and security issues will also have to be taken into account. Major policy initiatives designed to have a substantial impact will require global alliances. "We need leadership," Davis said, "and that leadership has to come with the support of people in democracies from the top." Policies also have to make sense in a highly competitive economic framework. Governments will not allow themselves to be left behind if policies damage their industries in the short term.

Meeting several immediate requirements can make planning for that period more effective, according to Davis. First, much higher resolution climate models are needed to predict and assess the impacts of climate change in order to weigh the real societal costs and benefits of alternative strategies. For the first time, governments are looking seriously at mitigation and adaptation strategies, Davis emphasized. As inputs for this planning, governments need much more detailed assessments of how climate will and could change and the consequences of change.

Second, much more analysis is needed regarding technologies expected to be available during the second half of the 21st century. Because relatively little is known now about such technologies, debates over how to reduce emissions by 50 or 80 percent by 2050 are taking place without a proper analytic framework. "What does it all look like?" asked Davis. "You can talk about bits and pieces. We need an integrated assessment of what those pathways are like."

Improving modeling capabilities will contribute to the development of strategies that involve energy technology choices and efforts to develop new technology options. In these efforts, independent advice from groups like the National Academies will be absolutely essential, according to Davis. Assessments of energy issues need to be fair, rigorous, and peer reviewed. Conclusions need to be strategically relevant and innovative. They need to generate options, clarify decision making, and ease the process of moving forward.

The nexus of energy and the environment poses an extraordinary challenge, said Davis. Responding to this challenge will require a societal transformation that will take place across generations. "These things never happen in straight lines," said Davis, "and they require immense courage. Not just of political leaders. That courage can come from anywhere, in any context."

3

The Developing World—The Case of China

Much of the increase in carbon dioxide emissions over the next few decades will come from the developing world. As James Schlesinger pointed out at the summit, the amount of coal burned in China is projected to increase from 2.3 billion metric tons to 4.5 billion metric tons annually by 2030. Similarly, India is planning to build a substantial number of new coal-fired power plants. Indonesia is shifting to coal as it seeks to increase its exports of oil. And in Russia, the energy company Gazprom recently bought the country's largest coal company so that it can increase the use of coal internally while exporting more natural gas.

All of these countries have immediate environmental problems linked to energy use that are currently more pressing than the eventual prospect of substantial climate change. "[China's] rivers are poisoned, and its poisoned rivers flow into Russia, which does not make the Russians happy," said Schlesinger. "China's air pollution is abysmal. It hasn't cut sulfur oxides. It hasn't cut nitrogen oxides as much as it should. A country that has those kinds of pollution problems is not going to have concern about the release of greenhouse gases very high on its priority list."

This chapter focuses on China as emblematic of the issues surrounding energy and the environment in the developing world. China was the developing nation discussed most extensively at the summit, and the magnitude of the problems it faces is greater for China than elsewhere. But observations made regarding China apply in many other developing countries as well, even if at a smaller scale and with somewhat different circumstances.

CHINA'S ENERGY USE IN A GLOBAL CONTEXT

Kelly Sims Gallagher began her talk at the summit by comparing China's energy use with that of the United States (Table 3.1). In 2006, China consumed 72 percent as much energy as did the United States. However, China is quickly catching up. Total energy consumption in China increased 70 percent between 2000 and 2005. And total coal consumption increased by 75 percent during the same time period. However, because China's population is more than four times that of the United States, the country's per capita use of energy is much lower.

China's oil imports also have grown rapidly, but they are still less than one-third U.S. levels. China's coal consumption, however, is already twice that of the United States. Coal is China's most abundant energy resource, even though the United States has much greater reserves. Coal accounts for 93 percent of China's remaining fossil fuel resources. Three-quarters of China's electricity comes from coal—526 gigawatts, as of 2007. Hydropower accounts

TABLE 3.1 Energy Comparisons Between the United States and China

	USA	China	China/ USA
Total energy consumption, 2006 (million tons of oil equivalent)	2,326	1,697	72%
Net oil imports, 2006 (million barrels per day)	12.3	3.4	28%
Total oil consumption, 2006 (million barrels per day)	20.6	7.4	36%
Electricity capacity, 2007 (gigawatts)	992	713	63%
Coal consumption, 2006 (million tons of oil equivalent)	567	1,193	210%
Reserves (percent of world)			
Coal	27%	13%	
Oil	2%	1%	
Gas	3%	1%	
Passenger cars (cars, pickups, SUVs), 2007 (millions)	~230	~30	13%
Total carbon dioxide emissions, 2007 (billion metric tons)	~6	~6	100%
Population	300 million	1.3 billion	433%

SOURCE: Kelly Sims Gallagher, "The Rise of China," presentation at the Summit on America's Energy Future on March 13, 2008.

for 20 percent of its electrical capacity, with nuclear, wind power, and other sources playing much smaller roles.

In 2006, China installed 101 gigawatts of new power. Ninety of those gigawatts came from coal-fired power plants. In 2007, China installed an additional 91 gigawatts of new power. "To put these astonishing numbers into perspective, Germany's entire electricity generation system as of 2005 was 125 gigawatts," said Gallagher. "India's was somewhat more, but close to Germany's level. So in 2 years China has built the equivalent of India and Germany's electricity capacities combined."

Coal is such a dominant source of energy in China that the country's current mix of electrical energy sources probably cannot be altered anytime soon, Gallagher said. Natural gas is not commonly used for power in China because of its high price and lack of availability. China is aggressively pursuing alternative sources of energy, such as solar hot water and small hydropower, and is also building nuclear power plants. But the fraction of energy from alternative sources is still dwarfed by the energy coming from coal.

The United States has approximately 230 million cars, light trucks, sport utility vehicles, and vans, whereas China has approximately 30 million. However, the number of vehicles in China is rising dramatically. Partly as a result, China's consumption of oil products has gone up substantially in recent years. Increased use has contributed to the recent increase in global crude oil prices. "Of course, the rise in world crude oil prices cannot be attributed only to demand growth in China," Gallagher emphasized, "but I do believe it's been a significant factor."

The same thing is happening with coal, Gallagher pointed out. In the first 6 months of 2007, China imported more coal than it exported for the first time in history. China's consumption of coal grew by 9 percent in 2007, which means that China's coal consumption could double from 2007 to 2015 if growth continues at that pace. In recent years, China has been accounting for most of the growth in the world's coal consumption. And as with oil, China's increased demand for coal has been correlated with rising prices. Also, the Chinese government's leading energy group projected that the country will need 384 gigawatts of new coal-fired power by the year 2020.

CHINA'S CONTRIBUTION TO CLIMATE CHANGE

Today, total carbon dioxide emissions from the United States and from China are approximately equal. However, as with energy use in general, per capita emissions are much lower in China than in the United States. Also, to some extent, energy consumption is increasing in China as it manufactures more products for export to other countries, which can have the effect of transferring emissions from those countries to China.

Turning around the current increase in global greenhouse gas emissions over the next two decades is a tremendous challenge, Gallagher pointed out. According to the IEA, carbon dioxide emissions from fossil fuels are projected to grow by approximately 50 percent by 2030 without active policy intervention (Figure 3.1). At that point, China and the United States will together account for about 45 percent of greenhouse gas emissions. For those two countries to be on a declining path (indicated by the arrow on Figure 3.1), substantial changes in each nation will be necessary. "I don't think it is possible [in China] without aggressive, sustained, and determined action," said Gallagher. "I'm optimistic it could be done, but early action and considerable financial resources will be needed. Time is not our friend here."

Electricity generation accounts for more than half of China's greenhouse gas emissions, and three-quarters of that electricity goes to industry rather than households. Iron, steel, and cement production together leads to more than a third of China's greenhouse gas emissions. Reducing greenhouse gas emissions in China therefore has a direct impact on its economy. Today, coal accounts for 80 percent of China's greenhouse gas emissions, and that percentage is projected to remain unchanged through 2030 in a business-as-usual scenario. In the United States, coal accounts for 35 percent of emissions and is projected to account for 39 percent by 2030.

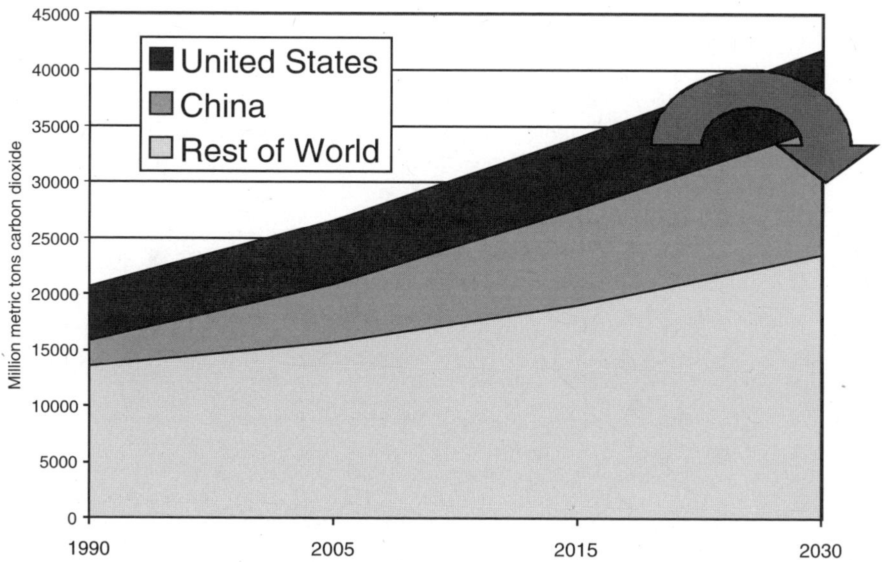

FIGURE 3.1 Projections of carbon dioxide emissions through 2030. Indicated by the arrow is the turn toward emissions reductions needed to help stem climate change. SOURCE: Kelly Sims Gallagher, "The Rise of China," presentation at the Summit on America's Energy Future on March 13, 2008; based on data from IEA (2007).

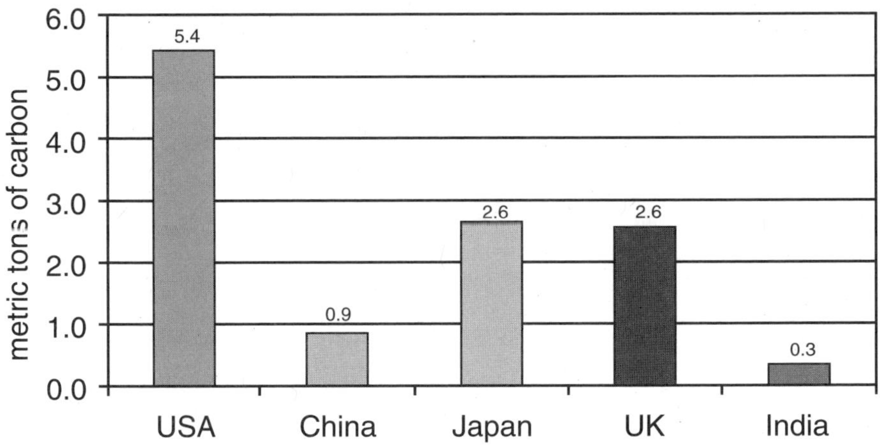

FIGURE 3.2 Per capita carbon emissions from fossil fuel burning in 2003 were much higher in the United States than in other countries. SOURCE: Kelly Sims Gallagher, "The Rise of China," presentation at the Summit on America's Energy Future on March 13, 2008; based on data from Oak Ridge National Laboratory, Carbon Dioxide Information Analysis Center.

In China emissions from oil for transportation account for only 6 percent of emissions, whereas in the United States, transportation is responsible for 43 percent of emissions—seven times China's transport emissions. Emissions from vehicles will grow in China by 2030, but the increased emissions from coal will dwarf those from vehicles. In fact, if every vehicle in the United States instantly ceased operation, the growth in emissions from China's coal would fill the gap in a few months, said Gallagher.

Issues of fairness and equity loom large in any discussions of greenhouse gas emissions in the United States and China. People in the United States still emit much more carbon dioxide per capita than do people in China (Figure 3.2). And millions of Chinese continue to live in poverty. According to the World Bank, 135 million Chinese live on less than $1.00 per day, and millions more live just above that arbitrary poverty line. "This makes the equity issue quite complicated," said Gallagher.

POLICY INITIATIVES IN CHINA

China's energy-related challenges are "numerous, intractable, and very complicated," Gallagher said. The country needs energy to sustain economic growth. It is becoming increasingly reliant on foreign sources of oil and natural gas. It needs to provide modern forms of energy to China's poor. Its urban air pollution is increasingly severe. It is facing growing concerns domestically and internationally about global climate change. And it needs access to affordable advanced energy technologies to address all of these challenges.

China has already done a lot to address its energy challenges, Gallagher said. First, it has launched very aggressive programs to improve the efficiency of energy use. For example, China's fuel efficiency standards for automobiles are much stricter than are U.S. standards. Fuel efficiency standards that go into effect in 2008 call for automobiles to get approximately 35 miles per gallon, a goal that U.S. standards will not reach until 2020 (although methods of enforcing such standards are considerably less developed in China than in the United States). Automakers are not allowed to produce a vehicle until the model has been certified to meet the efficiency standard. In addition, in its eleventh 5-year plan the Chinese government has set a very ambitious target of a 20 percent reduction in the use of energy per capita by 2010. This goal is proving difficult to achieve, Gallagher said, but the goal nevertheless exists.

The incremental costs of new energy technologies are a big hurdle in China, according to Gallagher. She recently did a study with colleagues at the Chinese Academy of Sciences on the capital costs of advanced coal plants in China. Several different kinds of plants are very efficient and fairly clean in terms of conventional pollution control, and China has three such integrated gasification combined-cycle (IGCC) coal-fired plants under construction for demonstration purposes. IGCC plants also provide the option of capturing carbon dioxide at a reasonable cost. However, the capital costs of IGCC and other advanced coal-fired plants are considerably greater than the costs of conventional coal-fired plants. Given the projections of new coal-fired capacity needed in China between now and 2020, the additional cost of building IGCC plants instead of more conventional plants, so that the option would be available of later capturing and storing carbon dioxide, would be approximately $190 billion—or about $16 billion per year. "That sounds like a big number in some respects, and in some respects it doesn't," said Gallagher.

Gallagher pointed out that the Chinese government could take a number of steps to address the challenges it faces. First, it could slow down economic growth or population growth. Both options are difficult, but Gallagher noted that the government has recently sought to manage its economic growth better, and it has recently retained its one-child policy.

The government also could promote an economic shift to lighter industry. For example, a fair amount of carbon is emitted in producing items for exports, including items sent to the United States. Such a shift is already a stated goal of the Chinese government, Gallagher said.

China could move more aggressively toward lower-carbon fuels, Gallagher said. Russia has considerable natural gas, but China has been reluctant to rely heavily on that source. This could change in the future, according to Gallagher.

The government could allow energy prices to rise. Although useful, this approach would risk inflation and would be very regressive in terms of its effects on poor people in China, which is a point often emphasized by Premier Wen Jiabao.

China also could more aggressively pursue efficiency through environmental pollution standards. But current standards are already fairly aggressive, Gallagher said, and a more important issue is to better enforce the standards that are already in place.

China could strengthen the legal system to foster a culture of compliance and enforcement of laws and regulations. The nation also could make greater investments in technological innovation for low-carbon technologies.

Finally, China could allow environmental activism to flourish and respond to it. In 2006, Gallagher noted, the minister of the State Environmental Protection Administration announced that 51,000 environmental protests occurred in 2005 in China. Not all of those were really environmental protests, Gallagher said. Many were protests about issues that more directly affect the Chinese people, such as land being taken away from poor farmers and converted to other uses. "Still, it shows that there is a rising sense of urgency in China about environmental issues." And the same kinds of activism have been responsible for greater environmental consciousness and increased environmental legislation in the United States.

U.S.-CHINA COOPERATION

"The United States and China are clearly the two countries with the unique ability to make or break the climate change threat," Gallagher said. "If either one fails to effectively manage its greenhouse gas emissions during this century, it's really almost impossible to substantially reduce the threat of climate change. If both fail, the game is really over."

Because of China's heavy reliance on coal and its current state of economic development, it will be much harder for China to reduce its greenhouse emissions than it will be for the United States, according to Gallagher. China still lacks many of the necessary institutions, policies, and enforcement mechanisms needed to foster vibrant markets, technology transfer, and environmental protection, especially at the provincial and county levels. At this point, the Chinese government is still most concerned about energy security and economic growth. Although many in the government are aware of the environmental dimensions of energy use and are beginning to take steps to protect the environment, maintaining stable economic growth, which creates support for the current leadership, is the top priority.

The United States needs to accept the fact that it will have to help China reduce its emissions, Gallagher said. She proposed several options for how the two countries might work together to confront the climate change challenge.

First, once the United States has established a domestic mandatory program to reduce greenhouse gas emissions, it should ask China to adopt a program that is suited to its circumstances, Gallagher suggested.

The United States should consider forming a bilateral or multilateral investment fund to accelerate the deployment of low-carbon technologies in China, she continued. Such a fund could provide low- or no-interest loans, or even direct grants, for major new industrial facilities or power plants that use low-carbon technologies. The Chinese government also should pay into that fund. Such a fund would create a market for low-carbon technologies in China. It also would be advantageous for U.S. firms that have technologies to sell in China. This is not a politically popular idea, Gallagher admitted, "but I believe it is necessary."

Energy technology collaborations between the two countries could be greatly increased, Gallagher stated. Joint research, development, and demonstration projects would be valuable for both countries and could bring the U.S. private sector into better contact with Chinese partners. For example, demonstrations of carbon capture and storage, renewable energy, energy storage, and energy efficiency technologies can be greatly expanded.

Finally, Gallagher pointed out, the United States could significantly bolster cooperative activities that are aimed at the collection and reporting of data in China and at policymaking, institution building, and enforcement of environmental policies.

Government officials in China are very interested in cooperating with other nations on energy issues, Gallagher observed. But political issues erect high barriers to such cooperation. As Schlesinger pointed out, U.S. politicians have many differences with government leaders in China. There are concerns about China's military activities, its continual attempts to hack into U.S. defense computers, its record on stealing technology, Tibet, human rights, and women's rights. As Schlesinger said, "I would hesitate—and I'm not the most timid person in the world—to go to the United States Congress and ask them to finance a reduction in the release of greenhouse gases from the People's Republic of China."

Part II

Energy Supplies

A portfolio of options needs to be put in place, because our energy is going to come from a multitude of sources.

—Michael Ramage

4

Petroleum and Natural Gas

As the use of energy has risen, questions have multiplied about whether adequate supplies will be available to meet demand. The world's oil still comes largely from giant and supergiant oil fields that were discovered more than 50 years ago, James Schlesinger observed at the summit. Many of these fields are now going into decline, including the Burgan oil field in Kuwait, Canterell in Mexico, the North Sea, and the north slope of Alaska. The Saudis are trying to sustain production in Ghawar, the massive field that provides more than 6 percent of the world's oil, but sooner or later that field, too, will go into decline.

"We face a painful transition," said Schlesinger, "to a future in which we hit a limitation, a plateau, in the ability to produce crude oil." Schlesinger pointed out that the concept of "peak oil"—when production reaches a maximum and begins to decline—is drawn from geological analogies and ignores such things as technology and the impact of price rises. Nevertheless, supplies of petroleum will be increasingly constrained. EIA projections call for the production of conventional oil to rise from the current 86 million barrels a day to about 118 million barrels a day by 2030. "That means that we must find or develop, given the decline curve and higher aspirations, the equivalent of nine Saudi Arabia's. I think the probability of being that successful is very low," said Schlesinger.

Furthermore, with many oil and gas fields under the control of national oil companies, access to these resources is becoming more restricted. For example, Russia has reasserted its control over oil production, Schlesinger observed, and the Russians have made it clear that they are not trying to solve the world's energy problems. Other countries have taken a similar stance: they will develop their resources based on their own self-interests.

Similar questions surround supplies of natural gas. As Ernest Moniz pointed out, the United States increasingly has turned to natural gas for electricity production, partly because the capital costs of natural gas plants are lower than those for other energy sources and because natural gas produces lower levels of greenhouse gas emissions than do either petroleum or coal. But the increased use of natural gas has driven up its price. As a result, U.S. manufacturers that have depended on natural gas for direct energy conversion or as a feedstock are being driven overseas.

THE HARD TRUTHS ABOUT OIL AND GAS

Rod Nelson summarized the conclusions of a major study of the world's oil and natural gas supplies that was done by the National Petroleum Council (NPC), a federally chartered and privately funded advisory group that represents the oil and gas industries' views to the federal government (NPC, 2007). In 2005, Secretary of Energy Samuel Bodman asked the NPC to study whether global oil and natural gas supplies can keep pace with growing world demand. Key questions were, What does the future hold for global oil and natural gas supply? Can incremental oil and natural gas supply be brought on line, on time, and at a reasonable price to meet future demand without jeopardizing economic growth? What oil and gas supply strategies and demand-side strategies does the NPC recommend that the United States pursue to ensure greater economic stability and prosperity?

The NPC study, which examined the period between now and the year 2030, involved a large number of study groups, subgroups, and subcommittees working under the direction of the Committee on Oil and Gas. More than 350 people participated in these groups—with 65 percent coming from outside the oil and gas industry—and more than 1,000 other individuals and groups provided input to the study. The Committee on Oil and Gas developed what it called six "hard truths" about oil and gas (NPC, 2007). The first is that:

> *Coal, oil, and natural gas will remain indispensable to meeting total projected energy demand growth.* (p. 5)

As pointed out in Chapter 1, the IEA projects that global energy use will rise from about 450 quadrillion Btu in 2004 to about 700 quadrillion Btu in 2030 (IEA, 2007). The proportional contributions of energy sources do not seem to change much in the projections, Nelson observed, but the total use of energy grows dramatically (Figure 4.1). "The pie is getting bigger," he said, "by 50 percent. So, in fact, biomass, solar, and wind are growing. In some cases, [they are] tripling or quadrupling in this timeframe." Also, much of the future growth of energy use will occur in the developing world, which relies heavily

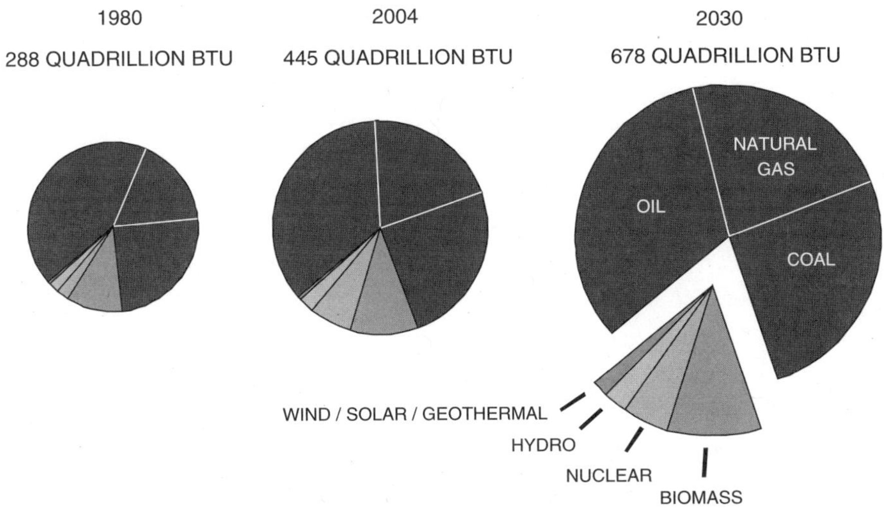

FIGURE 4.1 The use of energy worldwide is projected to increase 235 percent from 1980 to 2030. SOURCE: Rodney Nelson, National Petroleum Council, "Facing the Hard Truths About Energy," presentation at the Summit on America's Energy Future on March 13, 2008; data from International Energy Agency 2006 reference case (IEA, 2006).

on fossil fuels. So wind power in developed countries can increase dramatically without substantially changing its proportion in the total.

The second hard truth in the 2007 NPC study is that:

> *The world is not running out of energy resources, but there are accumulating risks to continuing expansion of oil and natural gas production from the conventional sources relied upon historically. These risks create significant challenges to meeting projected energy demand.* (p. 5)

Estimates of oil capacity from both conventional and unconventional sources, such as heavy oil or oil shales, remain large. Humans have used about 1.1 trillion barrels of oil since the dawn of the oil age, Nelson said. Conventional supplies of oil exceed 3 trillion barrels, and the use of unconventional sources of oil could add a substantial amount to that (Figure 4.2). "Quite a bit of oil [is] still left in the ground," said Nelson. "The more you explore, and the more you learn about the earth, the more you find."

However, oil production is the more relevant measure, Nelson acknowledged. The NPC study (NPC, 2007) looked at projections for 2030 made by different organizations, which ranged from 85 million barrels a day (approximately today's level) to 130 million barrels (Figure 4.3). "Reasonable people

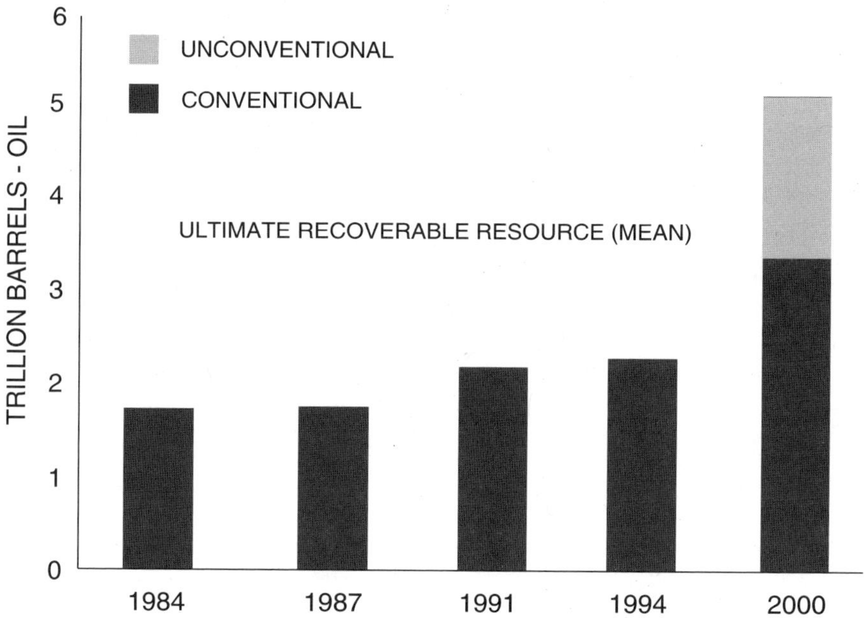

FIGURE 4.2 Estimates of the oil resource base exceed 3 trillion barrels for conventional supplies, with unconventional supplies (which were estimated for the first time in the year 2000) adding a more uncertain but potentially large amount to that base. SOURCE: U.S. Geological Survey (2000).

come up with different conclusions here," he said. "The people at the bottom of the projection spectrum are the people most worried about oil peaking sooner rather than later. . . . The people at the top end of the spectrum tend to be consulting organizations that are doing forecasts of oil supply." Assumptions in different projections include whether sufficient investments in infrastructure will occur, whether geopolitical events will reduce access to oil, and whether environmental issues will become more pressing. An aggregation of the data yielded a midpoint projection of about 100 million barrels per day.

The third hard truth in the 2007 NPC study involves energy sources:

> *To mitigate these risks, expansion of all economic energy sources will be required, including coal, nuclear, renewables, and unconventional oil and natural gas. Each of these sources faces significant challenges—including safety, environmental, political, or economic hurdles—and imposes infrastructure requirements for development and delivery.* (p. 5)

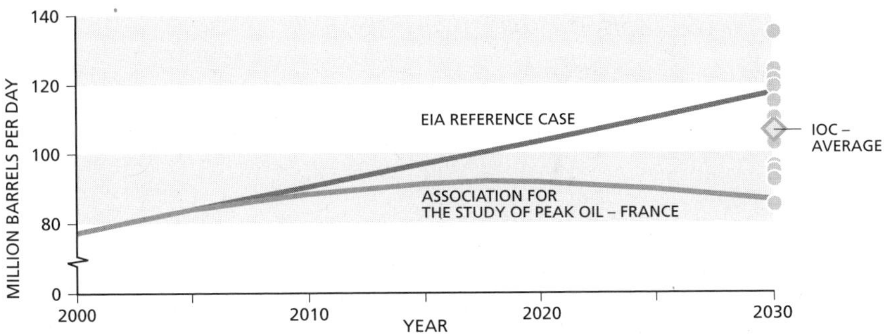

FIGURE 4.3 Projections of oil production range from a low of about 80 million barrels in 2030 to a high of more than 130 billion barrels. NOTE: Average of aggregated proprietary forecasts from international oil companies (IOC) responding to the National Petroleum Council (NPC) Survey of Global Energy Supply/Demand Outlooks (NPC Survey). For identification of other aggregations and outlooks shown here, see NPC (2007), Chapter 2, sections entitled "Analysis of Energy Outlooks" and "Global Total Liquids Production." Data from Energy Information Administration, *International Energy Outlook 2006*, Office of Integrated Analysis and Forecasting, U.S. Department of Energy, 2006, and the NPC Survey. SOURCE: NPC (2007; Figure ES-9). © National Petroleum Council 2007. Reprinted with permission.

The infrastructure requirements are particularly demanding, Nelson pointed out. "We have been living off an infrastructure surplus in the energy business that occurred in the late 1970s and early 1980s, and it's about used up." The electrical grid infrastructure, the transportation infrastructure, the coal production infrastructure, and other sources' production and supply systems require attention. "And the size and scale of this business is such that we're talking about a lot of money," said Nelson.

The 2007 study's fourth hard truth focuses on energy security:

> *"Energy independence" should not be confused with strengthening energy security. The concept of energy independence is not realistic in the foreseeable future, whereas U.S. energy security can be enhanced by moderating demand, expanding and diversifying domestic energy supplies, and strengthening global energy trade and investment. There can be no U.S. energy security without global energy security.* (pp. 5-6)

The United States is the world's largest oil consumer and the third largest oil producer on the planet. But the largest increases in the flow of oil over the next two decades are likely to involve countries in Europe and Asia (Figure 4.4). For that reason, the United States needs to remain closely involved in discussions of the oil trade, Nelson said.

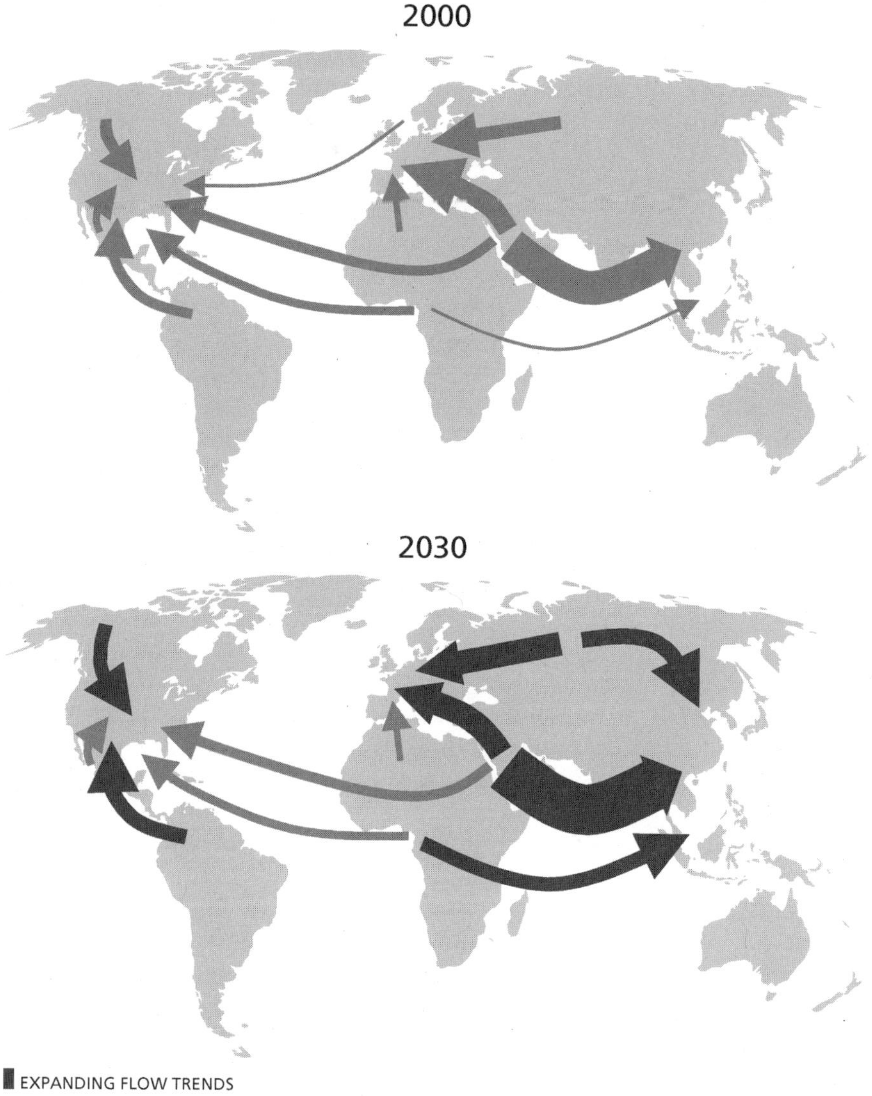

FIGURE 4.4 Expansion of the oil and natural gas trade will largely involve flows from the Middle East and Russia to other countries in Europe and Asia. SOURCE: NPC (2007; Figure ES-9). © National Petroleum Council 2007. Reprinted with permission.

The fifth hard truth from the 2007 NPC study involves the energy sector workforce:

> *A majority of the U.S. energy sector workforce, including skilled scientists and engineers, is eligible to retire within the next decade. The workforce must be replenished and trained.* (p. 6)

Since the early 1980s, relatively few young workers have entered the energy sector, Nelson pointed out, and relatively few people are in energy-oriented university programs in the United States (Figure 4.5). Making up the deficit is "not entirely possible from U.S. graduates."

The sixth hard truth identified in the 2007 NPC study deals with carbon emissions.

> *Policies aimed at curbing carbon dioxide emissions will alter the energy mix, increase energy-related costs, and require reductions in demand growth.* (p. 6)

This sixth NPC finding may not seem very dramatic given the current concerns about climate, Nelson said. "But I would remind you that this is the oil industry telling you this, and this was not, I would say, the generally held view 2 years ago."

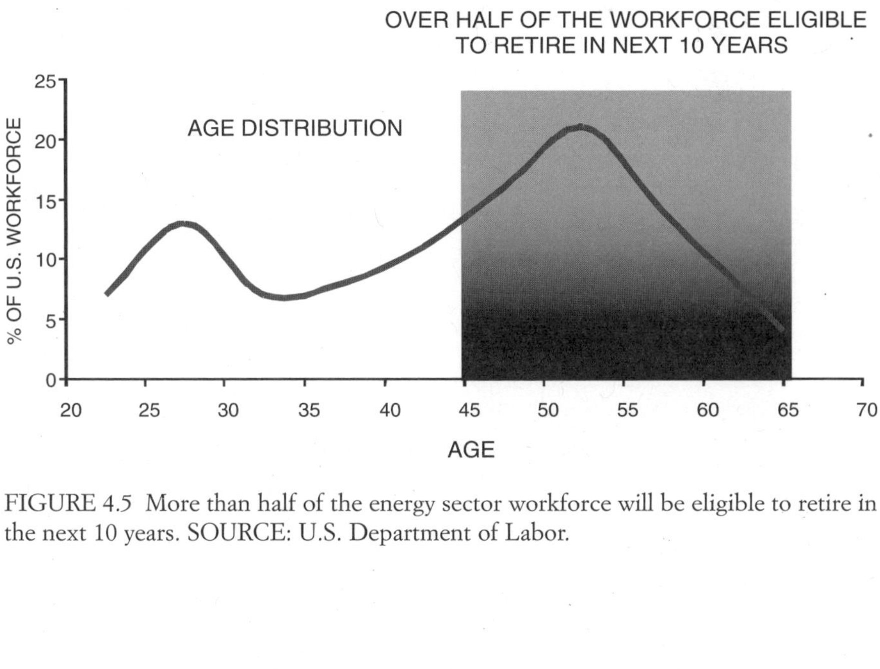

FIGURE 4.5 More than half of the energy sector workforce will be eligible to retire in the next 10 years. SOURCE: U.S. Department of Labor.

CONFRONTING THE HARD TRUTHS

The NPC's Committee on Oil and Gas laid out five core strategies for the United States to pursue (NPC, 2007):

- Moderate demand by increasing energy efficiency.
- Expand and diversify the U.S. energy supply.
- Strengthen global and U.S. energy security.
- Reinforce capabilities to meet new challenges.
- Address carbon constraints.

Each of these topics is discussed in parts III and IV of this report. Nelson's overall conclusion at the summit was that the challenge currently facing the United States regarding oil and natural gas is unprecedented. Meeting that challenge will require global efforts on multiple fronts, long time horizons, and major additional investments. There is no single easy solution, he said. Individuals, organizations, and governments need to begin taking action now and plan for a sustained commitment.

5

Coal

As Jeff Bingaman pointed out, the United States has more energy resources in coal reserves than the Middle East has in petroleum reserves. But the current methods for use of coal, either for electricity generation or for the production of liquid fuels, produce substantial amounts of carbon dioxide. For example, even if the conversion of coal to liquid fuels were 100 percent efficient, 1 ton of coal would yield about a half ton of fuel and 2 tons of carbon dioxide. The United States could "wind up spending a great deal of money on coal liquefaction plants that would then be rendered uneconomic in light of future developments related to global warming," said Bingaman.

Despite its environmental effects, coal use in the United States and other countries is currently on a rising trajectory. "Virtually any scenario that we see shows coal use growing," said Ernest Moniz. "It's cheap, abundant, and—in contrast to oil, for example—has a strong correlation between supply and demand." The three countries that use the most coal—China, India, and the United States—also are the three most populous countries in the world. Together they account for about 40 percent of the world's population and economic activity. Yet they use about 60 percent of the coal burned worldwide, and the amount of coal used in each country is increasing.

For coal to be a major source of energy in the future, much of the carbon it releases must be captured and sequestered underground, Moniz said. This carbon capture and sequestration (CCS) will require immense amounts of technology development. Also, CCS must prove to be economical in comparison with other technologies, including nuclear power or renewable energy sources. In contrast to the problems with nuclear waste, Moniz said, the challenge of CCS "is one where the experts are far more concerned than the public."

TAKING CARBON CAPTURE AND SEQUESTRATION TO SCALE

Moniz summarized the conclusions of a report on the future of coal that was recently conducted by a group at the Massachusetts Institute of Technology (Deutch and Moniz, 2007). According to that report, coal is today a cheaper source of energy than oil, natural gas, nuclear power, or renewable sources of energy. But the use of CCS technology to reduce future climate change will substantially increase the cost of coal as an energy supply. The MIT study set out to find a path that mitigates carbon dioxide emissions yet continues to use coal to meet urgent energy needs, especially in developing countries.

Maintaining and increasing the use of coal as a major energy source without harming the environment will require that tremendous amounts of carbon dioxide be sequestered, Moniz observed. A single coal-fired plant produces millions of metric tons of carbon dioxide per year, which translates into more than a billion barrels of carbon dioxide over the course of its lifetime. Mitigating climate risks will require that billions of tons of carbon dioxide be sequestered globally each year. No laws of physics rule out such an accomplishment, but achieving it will require, as Moniz put it, "exquisite reservoir management."

Carbon dioxide capture has been done before in refineries and other industrial settings. But those technologies have been extremely expensive. "We really need some new technology to improve cost and performance," Moniz said. Developing these technologies will require that many scientific and technological questions be addressed, including questions about the physics and management of underground reservoirs. Large investments in infrastructure also will be needed, and a broad range of regulations will need to be put in place dealing with such issues as permitting, liability, siting, and monitoring.

Once CCS technology is developed, economic incentives will be needed to spur its commercial application. The MIT study examined the effects of imposing a tax on the use of fossil fuels designed to encourage CCS and the development and use of other energy sources (Deutch and Moniz, 2007). The high-tax trajectory starts at $25 per metric ton of carbon dioxide in 2015 and increases at a real rate of 4 percent per year. The low-tax trajectory begins with a carbon dioxide emission price of $7 per metric ton in 2015 and increases at a rate of 5 percent thereafter.

Both taxes have a substantial effect on the amount of carbon dioxide released into the atmosphere (Figure 5.1). However, the high-tax scenario makes sequestration an economically attractive technology well in advance of the low-tax scenario (Figure 5.2). "If you start delaying projects for 10 years and then add 20 years for deployment, . . . the conclusion is [that we need] to begin the process now."

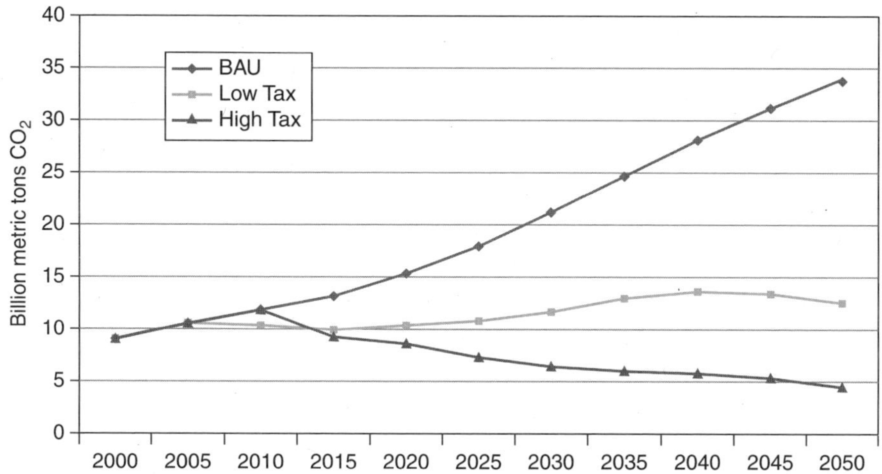

FIGURE 5.1 Global carbon dioxide emissions from coal would drop substantially from a business-as-usual (BAU) scenario through the imposition of taxes on carbon emissions. SOURCE: Deutch and Moniz (2007). Reprinted, with permission, from Ernest Moniz and Massachusetts Institute of Technology.

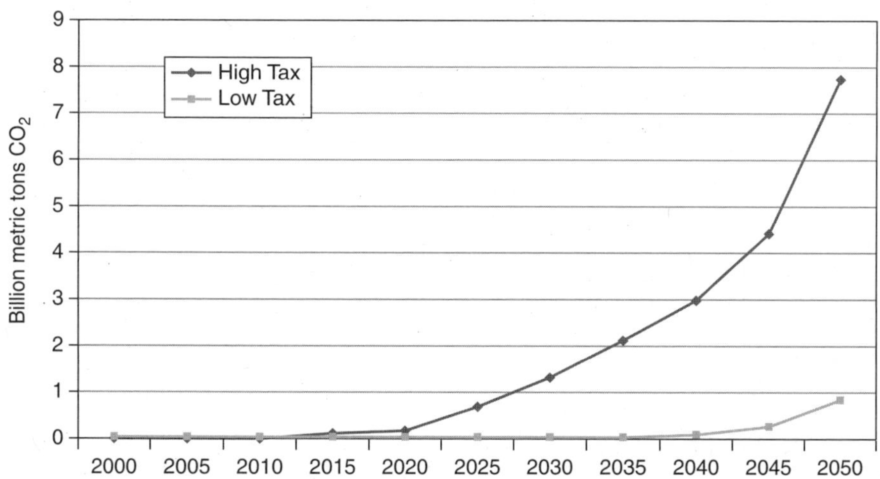

FIGURE 5.2 The annual sequestration of carbon dioxide, in billions of metric tons per year, would rise substantially with a high carbon tax and less substantially with a lower tax. SOURCE: Deutch and Moniz (2007). Reprinted, with permission, from Ernest Moniz and Massachusetts Institute of Technology.

MOVING FORWARD WITH DEMONSTRATION PROJECTS

To begin the process now requires that technology development and demonstration projects begin immediately. "We need to put a demonstration program in place over the next 10 to 15 years," said Moniz. "It must operate at large scale. It's not good enough to have a bunch of small projects."

The major problem is that large-scale demonstration projects are expensive—typically $100 million per year for a decade, "and that's significant change, even if you are a large oil company." Moniz called for roughly $4 billion of public funds over a decade for a portfolio of demonstration studies. Similarly, Steven Specker, in a summary of work done by the Electric Power Research Institute (EPRI), called for a series of pilot-scale projects involving various capture technologies. "We have to develop the pilots and focus on getting the cost of capturing carbon dioxide down," he said. "Then we have to scale those up to demonstrations." Finally, technologies need to be integrated into full-scale plants.

The adoption of CCS has important implications for the kinds of coal plants that are constructed in the future. Some kinds of plants are more easily adapted to CCS technologies than others, and some can be retrofitted much more economically if a decision is made later to adopt CCS. There is no clear technology winner at the moment, Moniz said, and different plants will be needed for different situations, such as different types of coal. "The real message is that we need several projects going on in parallel and not serially."

Specker laid out a timeline for the parallel development of different plant and sequestration technologies, noting that EPRI was recently involved in the startup of a pilot project in Wisconsin to capture carbon dioxide using chilled ammonia (Figure 5.3). "This is real hardware that's really going to break," Specker said. "It's really going to have problems. We're going to learn from it. We're going to figure it out. This is what it takes to get the technology evolved. Analysis doesn't do it. You have to build it. You have to operate it, you have to learn from it, and then you have to scale it up."

Both Specker and Moniz mentioned the recent cancellation by the Department of Energy of the FutureGen project, which was a $1 billion project to design, build, and operate a coal-fired power plant with CCS. Later in the summit, Samuel Bodman cited cost overruns for the decision along with a choice to spend the money on several projects rather than one. "We are not walking away from carbon sequestration," Bodman said. "On the contrary, we are going to fund it in a very aggressive fashion. . . . We're trying to redirect the money in a more intelligent way, but that's hard to do in Washington."

Moniz, in his talk, said that the reasons given by the Department of Energy for FutureGen's cancellation were that the demonstration projects needed to be closer to commercial application and that funding a portfolio of projects was a

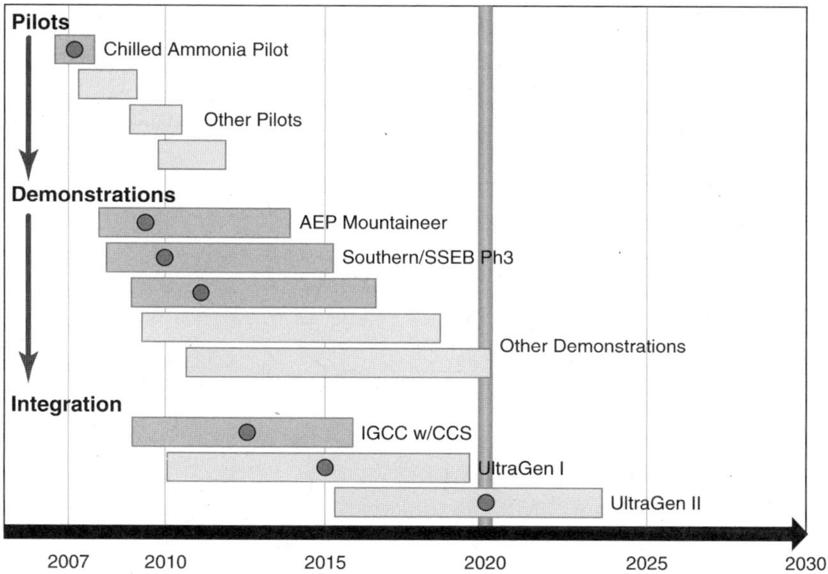

FIGURE 5.3 Advanced coal plants with carbon dioxide capture and sequestration have to be developed in parallel to be deployed by 2020. SOURCE: Energy Technology Assessment Center of the Electric Power Research Institute.

better option. "Both of those are good principles," Moniz said. "However, in our view, they are overwritten by the urgency of getting the race going. . . . We need to find a way of building on the work that has been done with FutureGen [while moving toward] a portfolio that emphasizes good commercial practice and multiple technology demonstrations." The highest priority at present, said Moniz, is to move aggressively to demonstrate sequestration at scale.

6

Nuclear Power

Nuclear power accounts for 20 percent of the U.S. electricity generated and does not release greenhouse gases into the atmosphere during power generation. Yet no new nuclear plants have been ordered in the United States for more than 30 years, Ernest Moniz pointed out, and the contributions of nuclear power to the nation's energy supply will decline unless a range of societal and economic issues are addressed. "For the short to medium term, said Moniz, the challenges "are frankly less technology and more policy and financing."

Moniz summarized the findings of a recent report on the future of nuclear power done by a group at the Massachusetts Institute of Technology (Deutch and Moniz, 2003). Three issues discussed in that report are critically important, he said: the economics of nuclear power, reprocessing and its connection to nuclear proliferation, and spent fuel management.

THE ECONOMICS OF NUCLEAR POWER

Like coal-fired power plants, nuclear power plants are capital intensive, Moniz observed. Furthermore, the price of large infrastructure projects has increased dramatically in the past few years—on the order of 75 percent over the past 3 years for large power plants. For that reason, although the 2003 report assigned a price to nuclear power of about $2,000 per kilowatt, current costs are substantially higher.

Nevertheless, companies continue to express renewed interest in the construction of nuclear power plants. The Nuclear Regulatory Commission has received 30 or so indications of interest from utility companies, with a number of those indications moving forward to more formal consideration. Most of these plants would be based on evolutionary improvements of existing designs with some improved safety features. Plants also could be built that incorporate advanced concepts developed through nuclear research and development programs.

The MIT report (Deutch and Moniz, 2003) argued for public support for a limited number of "first mover" power plants that represent safety-enhancing evolutionary reactor design. "If we want to demonstrate what the performance of these new plants will be technically and what their construction will look like in the new regulatory environment, . . . we need to get out there and build some plants," said Moniz. "You want to build a few of each design to establish the cost performance, the construction performance, and to [assess] the regulatory regime. Then it has to compete in the marketplace."

Another economic consideration involves the licensing of existing and future plants, Ray Orbach pointed out later in the summit. At a recent workshop on nuclear power, engineers in the nuclear industry were asked what their greatest problem was. Their response was, "cracks," Orbach said. Licensing of existing plants was extended from 40 to 60 years in the past. The issues associated with extending licenses to 80 years, which would substantially reduce carbon dioxide emissions, are now being examined. "We are now trying to extend the licensing [of nuclear plants] for 20 more years," said Orbach. But "there are real problems associated with fission energy, not the least of which are the materials issues surrounding the reactor itself." The Department of Energy is now funding research in materials science, nuclear physics, and advanced computing designed to understand and control processes that occur during nuclear power generation. For example, Orbach mentioned the possibility of developing self-healing materials that would reduce the problems observed in current reactors.

Steven Specker also emphasized that careful planning today can do much to extend the lifetimes of current and future generations of nuclear power plants. "It's like your own health," he said. "You better start taking care of yourself now. And we need to be doing things on today's plants that would allow [their lifetimes] to be extended."

REPROCESSING

As Orbach observed, the spent fuel that emerges from nuclear power plants still has a lot of energy left in it. By disposing of that fuel, the remaining energy is wasted. In addition, if the price of uranium increases, the energy left in spent fuel becomes even more valuable.

To extract this energy, the current administration has proposed a global nuclear energy partnership that would reprocess spent fuel in specialized reactors. These reprocessing technologies can reduce the amount of waste that needs to be managed and increase the amount of energy produced from a given quantity of uranium. However, Moniz and several other speakers at the summit were skeptical about the merits of reprocessing in the near-term future. First, reprocessing technology currently in use can be used in nuclear weapons. Second, Moniz and other speakers argued that the claims for the waste management benefits of reprocessing are exaggerated. John Holdren observed that reprocessing might reduce the volume of waste, but volume is not the constraint on the capacity of a waste repository. The constraint is the amount of heat generated by the waste, and that problem cannot be solved without reactors for reprocessing that are at least 40 to 50 years away. Reprocessing spent fuel makes nuclear energy "more complicated, more expensive, more proliferation prone, and more controversial," Holdren said. "If you want nuclear energy to be rapidly expandable, and to take a bite out of the climate change problem, you want to make it as cheap as possible, as simple as possible, as proliferation-resistant as possible, and as non-controversial as possible, and that means you don't want to reprocess any time soon."

At the same time, all of the speakers agreed that research on reprocessing for the longer term should be intensively explored. "We need to be investing in it," Holdren said, "but what we don't need to be doing is deploying reprocessing soon with technologies that are currently available because that will shoot nuclear energy in the foot." Moniz pointed out that far too little has been invested in advanced nuclear concepts, and "we are paying the price today for that lack of adequate research." For example, one possible approach would be for a balanced fuel cycle in which conventional reactors in "user" states feed spent fuel into a complex of advanced reactors located in "supplier" states (Figure 6.1). The user states would be assured of nuclear fuel supplies so long as spent fuel is returned to the supplier states. In this way, small nuclear programs could lease their fuel from states with advanced reactors, which would address proliferation concerns while concentrating and reducing the quantities of waste.

THE DISPOSAL OF SPENT FUEL

The management of spent fuel remains a difficult issue in the United States and around the world. Long-term geological isolation of spent fuel "appears to be scientifically sound in well-chosen sites with good project execution," Moniz said. Yet a system to dispose of nuclear waste has not yet been implemented anywhere in the world, and whether the designated U.S. site for spent fuel, Yucca Mountain in Nevada, can be licensed remains up in the air.

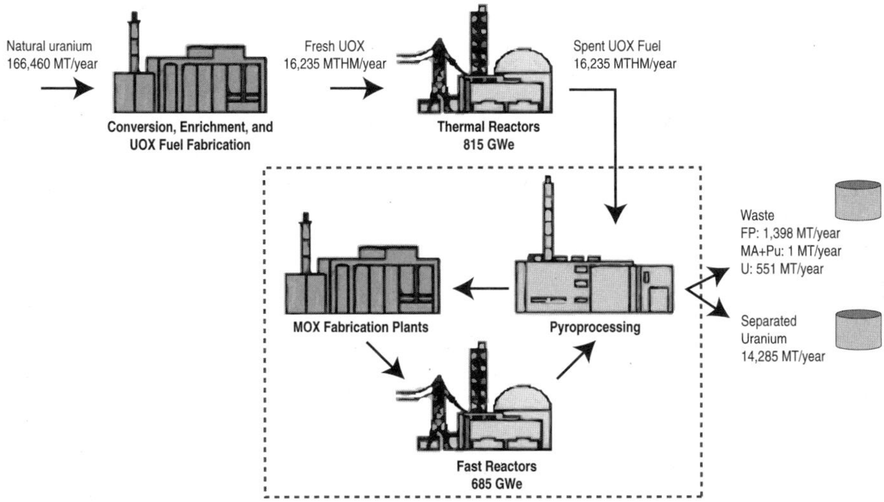

FIGURE 6.1 Under a closed fuel cycle plan, user states would send spent fuel to supplier states that would reprocess the fuel to produce waste and separated uranium for additional energy production. SOURCE: Deutch and Moniz (2003). Reprinted, with permission, from Ernest Moniz and Massachusetts Institute of Technology.

The MIT report concluded that storage of spent fuel for a century or so should be implemented as part of the nation's spent fuel management system (Deutch and Moniz, 2003). Ideally, spent fuel would be stored at centralized locations under federal control. Storage allows some of the heat of the fuel to dissipate, Moniz pointed out. It also would enable the further development of technologies and policies that could influence decisions about the management of spent fuel. "There is no urgent need for us to fill Yucca Mountain," he said. A "measured pace" is the better alternative. Interim storage in federal facilities would have the advantage of decoupling the private sector imperatives for running power plants from the longer-term and more difficult challenge of implementing and managing spent fuel disposal. However, there are political pressures to move forward to demonstrate that spent fuel can be well managed.

Interim storage also would provide more time for the large amounts of research that still need to be done on the disposal of nuclear waste, given that a substantial expansion of nuclear power generation will create much larger quantities of waste. As Orbach pointed out, if nuclear power is to provide a considerable portion of the future U.S. electrical power, "we would have to have eight Yucca Mountains by the end of this century in order to store the spent fuel."

FUSION ENERGY

Orbach also discussed fusion energy, which is "incredibly energy productive," he said. "But it takes place in the interior of stars, where the temperatures and pressures are a bit higher than those we have been able to achieve here on Earth."

Fusion reactors use isotopes of hydrogen as an energy source, including deuterium and tritium, and "there is enough deuterium in a body of water the size of Lake Erie to meet the energy needs of this earth for a thousand years," Orbach said. Fusion produces energetic particles and radiation that need to be captured in the wall of a reactor, which produces heat that can be used to generate electricity. It has been a very difficult process to master and cannot be mastered in the short term, but "we are entering a new era with ITER," Orbach said. ITER is an experimental fusion reactor in which hot gas is confined in a donut-shaped vessel and heated to more than 100 million degrees. The facility, which is sited in France and is a joint project of six nations and the European Union, is designed to produce about 10 times as much energy as it uses (Figure 6.2). The next step beyond ITER, Orbach said, will be a demonstration power plant based on fusion.

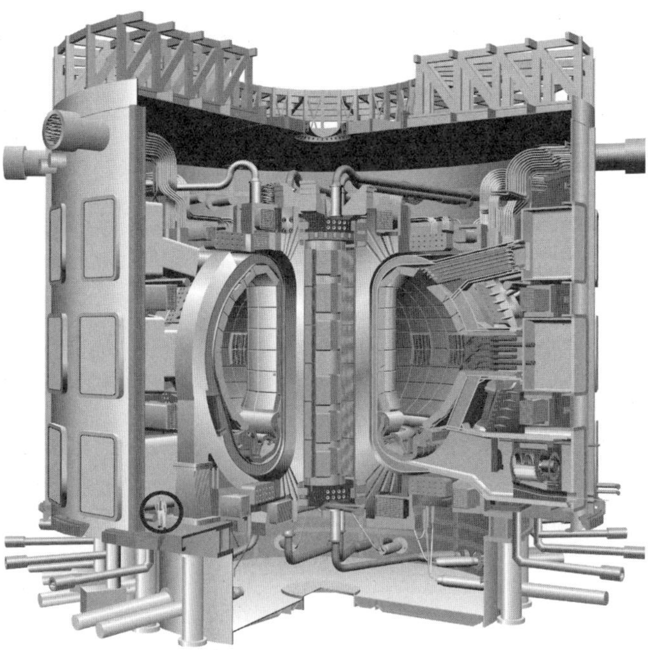

FIGURE 6.2 The fusion reactor ITER is designed to produce 10 times as much energy as it consumes. (Note size of human figure circled at lower left.) SOURCE: U.S. ITER Project Office, Oak Ridge National Laboratory.

7

Biofuels

When automobiles started to be used about a century ago, their developers seriously considered fueling them with ethanol, José Goldemberg pointed out at the summit. In fact, racing cars of the time were fueled with ethanol because the octane number for ethanol is better than that for gasoline.

Ethanol continues to have advantages over gasoline as a transportation fuel. It is an agricultural product that can be continually renewed. It does not emit impurities such as sulfur oxides and particulates, which, Goldemberg suggested, are a greater cause of concern than global warming in the large metropolitan areas of the developing world. If the proper feedstock and agricultural practices are used, the use of ethanol produces fewer greenhouse gases than the use of gasoline. Yet ethanol, when produced from food crops such as corn, also has serious drawbacks as an energy source, which requires that a careful assessment be made of the potential of biofuels to contribute to future energy supplies.

ETHANOL PRODUCTION IN THE UNITED STATES AND BRAZIL

The United States and Brazil are the main producers of ethanol in the world. Production in Brazil in 2006 was 17.8 billion liters. The area used to grow sugarcane to convert to ethanol was 2.9 million hectares, out of a total sugarcane production area of 5.0 million hectares in Brazil and 20 million hectares worldwide.

Ethanol production in the United States in 2006 was 18.4 billion liters. The area used to grow corn to convert to ethanol was 5.1 million hectares, out of a total area for corn production of 29 million hectares and 144 million hectares globally.

Together, an area of 8 million hectares was devoted to ethanol production in Brazil and the United States. The total area used for agriculture in the world is about 1,300 million hectares, so less than 1 percent of this area is being used for ethanol production, Goldemberg pointed out. Similarly, the total amount of ethanol produced by Brazil and the United States in 2006 was about 36 billion liters, which is less than 1 percent of petroleum use.

"You sometimes wonder why people are concerned so much," Goldemberg observed. However, there are several reasons for that concern, he added. Ethanol production in the United States and Brazil is slated to increase. In Brazil, production is expected to double by the year 2015. "That's not an extrapolation," Goldemberg said, "it's a calculation based on the number of [ethanol] plants that have been licensed and are under construction." In the United States, the 2007 Energy Independence and Security Act places an upper limit on corn-based ethanol of 56.8 billion liters per year by 2022, which is approximately a tripling from current levels. Furthermore, using more advanced cellulosic-based technologies, ethanol production in the United States could increase by at least another 80 billion liters per year by 2022, and in the European Union, where sugar beets are currently the crop used most often for ethanol, production could increase to 15 billion liters per year by 2020.

At that point, ethanol could replace 6 percent of the gasoline used in the world. Production at that level might enable the ethanol-producing companies to establish "a new OPEC of ethanol," Goldemberg said. "Saudi Arabia controls 12 percent of the oil, but it has a tremendous weight on what happens in the world. So this is not an insignificant matter."

Many countries have established mandates that call for particular levels of ethanol consumption in the future. Yet production costs vary greatly from country to country, from more than €50 per 1,000 liters for sugar beets in Germany to less than €15 per 1,000 liters for sugarcane in Brazil.

In addition, the amount of energy it takes to produce a given quantity of ethanol varies greatly for different crops (Figure 7.1). In Brazil, the extraction of the juice from sugarcane leaves considerable biomass, which is known as bagasse. This bagasse can provide all of the energy for the heat and electricity needed to produce ethanol. But cobs of corn do not have that same energy content, Goldemberg noted. As a result, fossil fuels need to be burned to produce ethanol from corn in the United States, making ethanol less attractive as a fuel in this context. Further, as the prices of fossil fuels rise, so will the cost of ethanol.

A concern unique to Brazil is the contention that the production of ethanol is causing the Amazon forest to be destroyed. But Goldemberg argued that this concern is misplaced. Most of the ethanol distilleries are in the southeastern

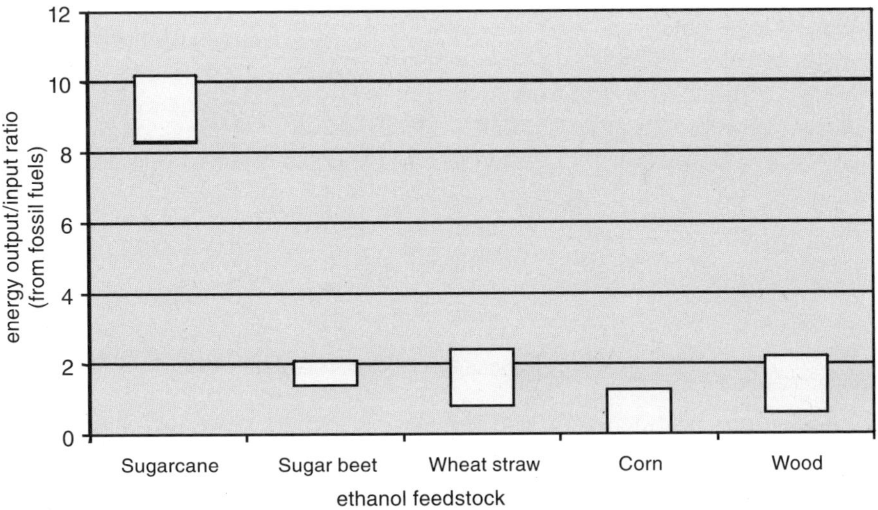

FIGURE 7.1 Feedstocks produce varying amounts of energy compared to energy inputs. SOURCE: José Goldemberg, State of São Paulo, Brazil, "Biofuels: How Much, How Fast, and How Difficult?," presentation at the Summit on America's Energy Future, March 13, 2008; based on data from Macedo et al. (2004), UK DTI (2003), and USDA (1995).

part of the country, with some in the northeast (Figure 7.2). Two-thirds of the ethanol in Brazil is produced in the state of São Paulo, which is far from the Amazon. There, sugarcane production replaced earlier crops such as coffee in response to government incentives to reduce the amount of petroleum imported into Brazil. "The Amazon Forest is being cut and no one more than the Brazilians—many Brazilians, including myself—are very annoyed at that, and we are fighting very strongly to eliminate that. But the deforestation of the Amazon is 1 million hectares per year. It's not 7, it's not 5, it's 1."

The Brazilian government's initial mandates in the 1970s called for 20 percent of gasoline to be replaced by ethanol. These mandates were "absolutely essential," Goldemberg said, "because as soon as you had a mandate the private sector had a sure market and a stimulus to develop the technologies." In addition, the government encouraged car manufacturers, which at that point were foreign, to produce cars that would operate on 100 percent ethanol. That created a problem, because with any agricultural product shortages and surpluses can develop. More recently, the problem has been addressed with the development of flex fuel cars, which can use different mixes of ethanol and gasoline. Today all gasoline is blended with some quantity of ethanol, and gasohol is economically competitive with gasoline in Brazil.

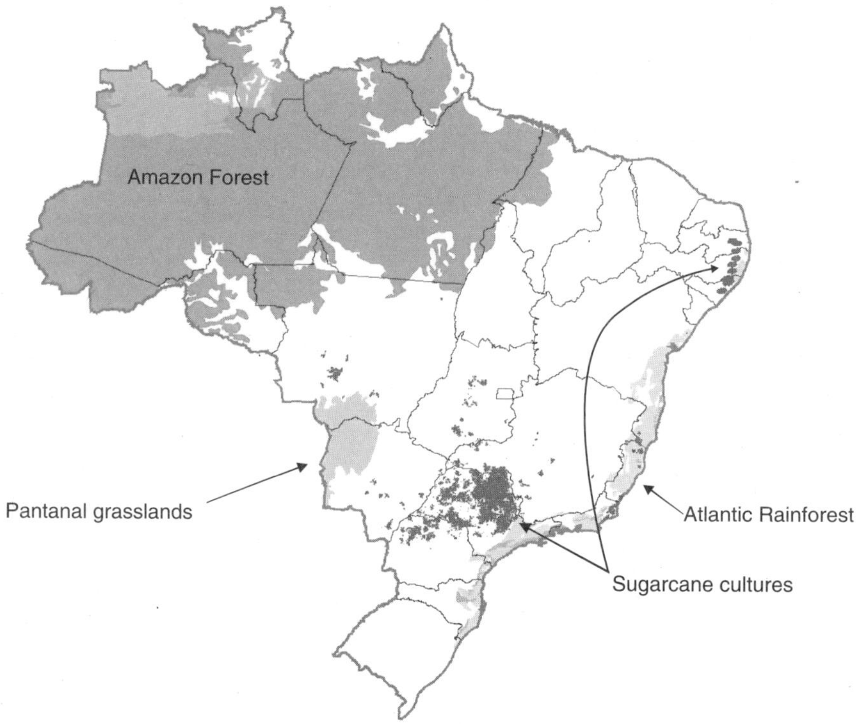

FIGURE 7.2 Sugarcane cultures are located in the southeast and northeast of Brazil, far from the Amazon Forest. SOURCE: Adapted from UNICA (2005; Figure 5, p. 131).

Since 1980, ethanol prices in Brazil have generally fallen, while the international price of gasoline (in Rotterdam) and the price of gasoline in Brazil have gradually risen, with a strong spike in recent years (Figure 7.3). "Ethanol was very expensive in the beginning," Goldemberg said. "As time went by, the learning curve decreased tremendously the cost of production."

Further increases in productivity can be expected in the future. From 1975 to 2005, the yield of ethanol per hectare in Brazil grew from 2,204 to 5,917—an annual increase of 3.77 percent—with most of the gains from agricultural improvements rather than distilling innovations (Figure 7.4). "It's a fantastic situation," said Goldemberg. "I wish all technologies would behave this way."

Now researchers are looking at the possibility of genetically modifying agricultural crops to further increase yields. Goldemberg cited informal information that additional gains of 30 percent are feasible in the near future.

In the United States, corn yields also rose from 1975 to 2005 (Figure 7.5).

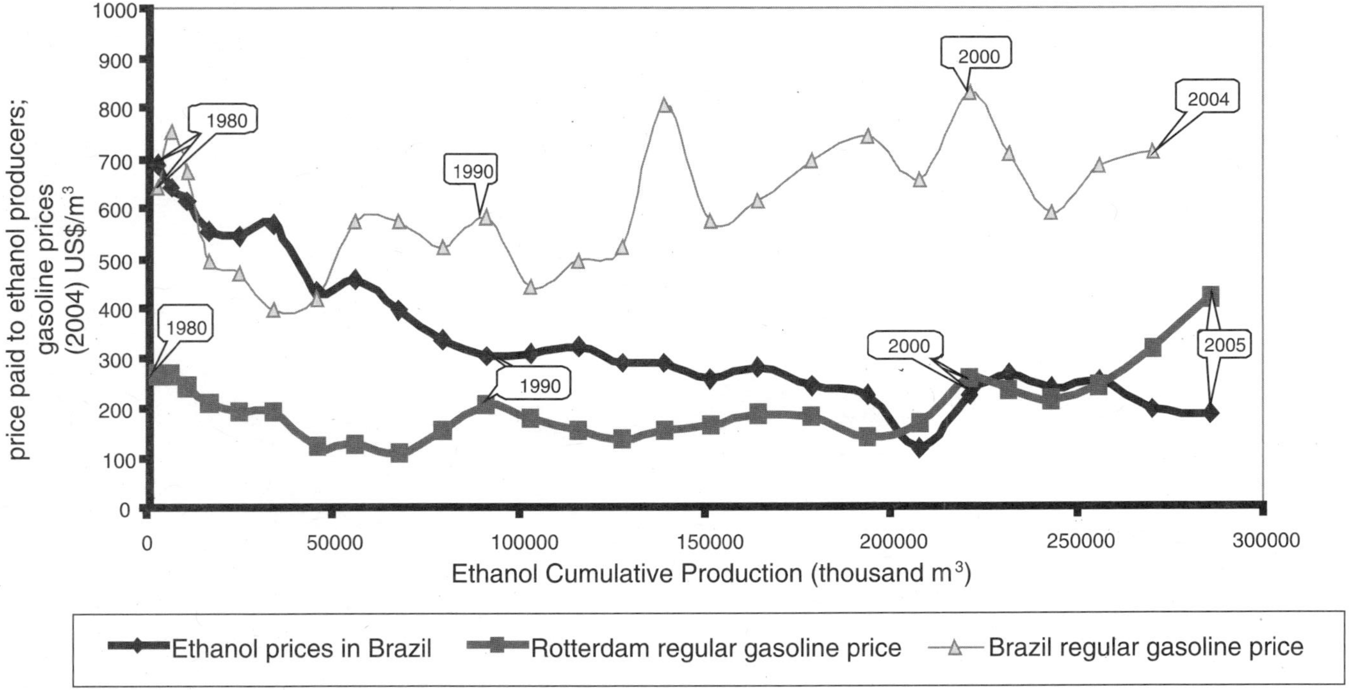

FIGURE 7.3 The price of ethanol has dropped steadily compared with the price of gasoline on the international market and in Brazil. SOURCE: José Goldemberg, State of São Paulo, Brazil, "Biofuels: How Much, How Fast, and How Difficult?," presentation at the Summit on America's Energy Future, March 13, 2008; based on Goldemberg et al. (2003).

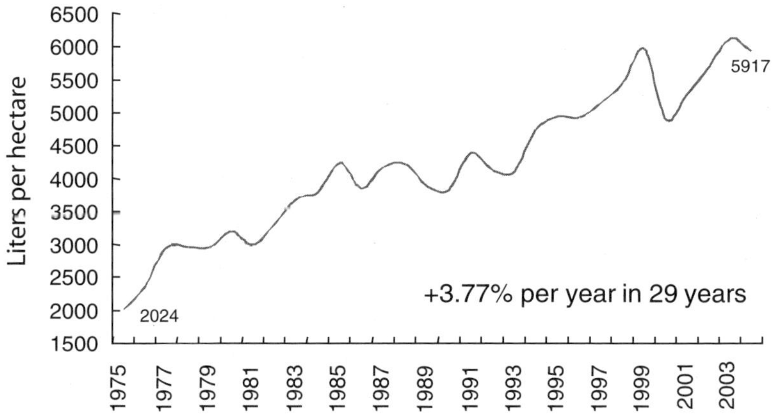

FIGURE 7.4 The yield (in liters) of alcohol per hectare has tripled over the past three decades. SOURCE: José Goldemberg, State of São Paulo, Brazil, "Biofuels: How Much, How Fast, and How Difficult?," presentation at the Summit on America's Energy Future, March 13, 2008; based on Datagro calculations in *The Brazilian Sugar Cane Agroindustrial Complex—Analysis of Status and Opportunities*, an unpublished study prepared by the Brazilian Reference Center on Biomass, 2007.

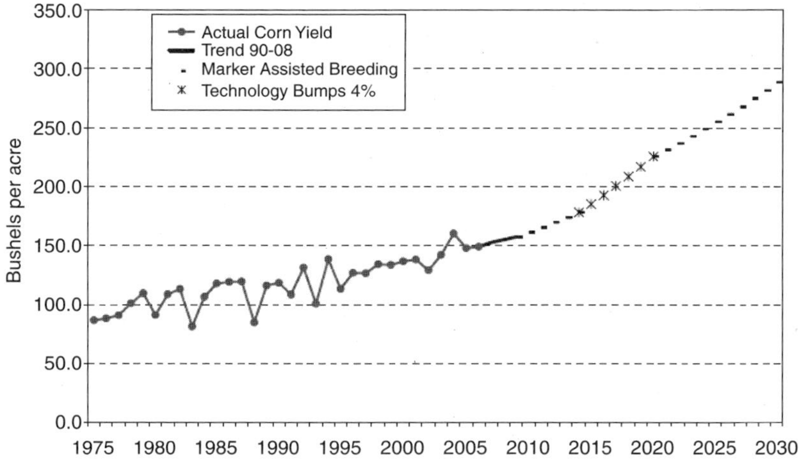

FIGURE 7.5 The yield (in bushels per acre) of corn could increase 4 percent per year during the period 2015 to 2020 through the application of new technologies. SOURCE: José Goldemberg, State of São Paulo, Brazil, "Biofuels: How Much, How Fast, and How Difficult?," presentation at the Summit on America's Energy Future, March 13, 2008; data from U.S. Department of Agriculture.

With a new genetic technology known as marker-assisted breeding, future gains could approach 4 percent per year. Such gains could reduce the conflict between ethanol and food production, Goldemberg said, which has become a major issue in the United States.

SECOND-GENERATION TECHNOLOGIES

Ethanol production currently is based on the long-established technology of fermenting sugars to produce alcohol. But new technologies could greatly increase the production of ethanol and other biofuels from agricultural products. For example, only about a third of the energy in sugarcane is contained in sucrose, Goldemberg noted. The remainder is contained in the bagasse and in the plant's tops and leaves. The bagasse consists largely of cellulose and hemicellulose. If cellulose could be converted into biofuels, the gain from all agricultural products, including other crops, grasses, and wood, could be increased considerably.

At the moment, the U.S. biofuels program is focused on ethanol, partly because "it's the only game in town," said Samuel Bodman. Also, money spent on ethanol goes to U.S. farmers rather than to some other country. But extracting ethanol from cellulose would have many advantages, Bodman said. First, it could cut greenhouse gas emissions by 80 percent compared with the use of fossil fuels. It also could be the first step toward an even greater potential breakthrough: making straight-chain hydrocarbons and aromatics—essentially equivalent to gasoline—from agricultural products. "That will change the nature of the business, if in fact it proves to be correct," said Bodman. "There's a lot going on, and we're at a very early stage in the evolution of this matter."

Going from cellulose to gasoline or diesel first requires getting at the cellulose in a plant, Ray Orbach said. Plant stems and stalks have great strength because of a polymer called lignin that surrounds the cellulose and gives the plant strength and protection. Lignin keeps enzymes from reaching the cellulose to break it down into sugars that can be transformed into fuel. Currently, high temperatures or strong acids are used to break down these materials. However, "termites do it every day," Orbach observed. In parts of the San Diego area, a stick can be pounded into the ground in the evening, and by the next morning Formosa termites will have digested the stick to extract nutrients. "Is there a way that we can figure out what the termite does, or how a cow's inner stomach works to break down plant fiber? Nature has figured it out."

The Energy Department recently conducted a competition to fund centers focused on advanced biofuels concepts. Three proposals were funded at a level of $10 million in 2007, with $25 million slated for the centers over the next 5 years. The Joint BioEnergy Institute led by Lawrence Berkeley Laboratory with five partnering institutes is using model organisms to search for breakthroughs in basic science and is exploring the microbial-based synthesis of fuels beyond ethanol. The Great Lakes Bioenergy Research Center led by the University of Wisconsin-Madison and Michigan State University with six partners is exploring the breakdown of plant fibers, methods to increase production of starches and oils (which can be more easily converted to fuels), and the environmental and socioeconomic implications of moving to a biofuels economy. The BioEnergy Science Center led by Oak Ridge National Laboratory with nine partners is focusing on the decomposition of plant fiber and on the potential energy crops switchgrass and poplars.

Already, this work has begun to produce dividends, according to Steven Chu. For instance, researchers at the Joint BioEnergy Institute have sequenced the DNA of the more than 100 microbes within a termite that help break down wood, creating the possibility that a designer microbe could be genetically engineered that has the proper combination of enzymes to break down wood in an industrial setting to generate fuels. Organisms also could be created to produce long linear hydrocarbon chains rather than ethanol. "In the first half-year of the DOE biofuels program, we now have our first gasoline-like and diesel-like fuels being generated by organisms," said Chu. Once such a technology is scaled up, the same process could be used to make gasoline, jet fuels, and diesel fuels. "It's not crazy to think of that," said Chu. "It's already proved in principle."

Some of the plants used for biofuels can be grown in marginal areas, such as salty ground. Also, plants especially suited to biofuels can be tremendously productive. For example, Chu showed a slide of miscanthus that grew almost 20 feet high from one fall to the next spring, with no applications of fertilizer or water (Figure 7.6).

The goal of federal research and development is to develop a sustainable carbon-neutral biofuels economy that meets more than 30 percent of the U.S. transportation demand without competing with food, feed, or export demands.[1] Orbach acknowledged that 30 percent is "a huge fraction. There's a lot riding on bioenergy, and we are up to our ears in trying to get it developed." In particular, environmental issues associated with bioenergy derived from plants must be addressed, including the effects on water, soil quality, land use, and biodiversity, Orbach said.

LAND USE

The use of land for biofuels production competes with the use of that land for other purposes, which has raised the concern that biofuels production will add to the upward pressure on food prices. But Goldemberg pointed out that the price of food over the long term has been declining gradually since the early 1970s, although food prices do tend to undergo substantial fluctuations.

---

[1]Officially, the DOE Biofuels Program goal is to displace 30 percent of gasoline consumption with biofuels by 2030 and to make cellulosic ethanol cost-competitive by 2012.

FIGURE 7.6 The feedstock grass *Miscanthus* in a non-fertilized non-irrigated test field at the University of Illinois yielded 15 times more ethanol per acre than did corn. NOTE: Estimates of the ratio for energy output to energy input for *Miscanthus* range from 12 to 19. SOURCE: Photo courtesy of Steve Long, University of Illinois at Urbana-Champaign.

Biofuels production will occupy "a relatively small amount of land," he said. Ethanol production also could have a stabilizing effect on oil prices, which would ease fluctuations in food prices. And the overall environmental consequences of ethanol production are small compared to the use of 86 million barrels of oil per day.

Goldemberg also looked at the prospects for increased sugarcane production in other countries. Brazil is not unique in its capacity to grow sugarcane. Approximately 100 countries in the developing world produce sugarcane, including all of the Caribbean countries. Goldemberg strongly urged U.S. policymakers to consider supporting the production and export of ethanol from the Caribbean, where the United States has a strong presence. "You might say, 'We'll be changing dependence on the Middle East for another dependence, on the Caribbean.' Well, the Middle East and the Caribbean are very different places."

At the height of the Roman Empire, the Roman legions refused to fight if they were not supplied with 1 kilogram of bread per day. The wheat for this bread did not come from Italy, Goldemberg pointed out. It came from North Africa, which became a rich area because of its food exports. Goldemberg optimistically concluded, "If the Roman Empire imported wheat to meet the needs of the Romans, I don't see why this is not considered a viable strategy for the United States."

8

Other Renewable Sources of Energy

Beyond fossil fuels, nuclear power, and biofuels are a variety of energy sources that will be part of the future portfolio, including solar power, wind power, hydropower, power from such sources as tides and waves, other forms of biopower, and hydrogen derived from these sources. Although most of these technologies were not much discussed at the summit, some received attention, and some more tentative sources of energy, such as enhanced geothermal energy, also were explored.

SOLAR POWER AND WIND POWER

Hydropower, which is a fairly mature energy source in the United States, was discussed very little at the summit. But several speakers touched on the prospects for greatly expanding the use of solar power and wind power.

As Ray Orbach pointed out, more energy from sunlight strikes Earth in an hour than all of the energy consumed on the planet in a year. Yet less than a tenth of 1 percent of our primary energy is derived directly from sunlight. "We have barely scratched the surface on solar energy, and the amount of energy available is so enormous that it is to our advantage to pay particular attention," Orbach said.

There are several ways of generating electricity from sunlight. One is to concentrate solar energy and use it to heat a liquid that can drive a turbine-generator set. This is the technology being pursued at a full-scale pilot plant that the Department of Energy is helping to fund in Nevada. Other techniques involve converting solar energy into electricity or into fuels. The challenge, said Orbach, is to reduce the costs and increase the efficiency of the conversion process, which often requires understanding and controlling phenomena at the nanoscale level.

One drawback to the increased use of solar power and wind power is that they tend to be intermittent sources of energy, in that they can be generated only when the sun is shining or the wind is blowing. Taking full advantage of these technologies will require the development of both energy storage technologies and "smart grids" that can control the flow of electricity from energy producers to energy users.

Several energy storage technologies are being developed. Although current batteries are not suited for large-scale storage, advanced batteries and electrochemical capacitors offer potential for the future. An older technology that remains useful is to pump water to a higher level and subsequently use the stored potential energy to generate electricity. Another such technology is compressed air storage, in which air is pumped into underground caverns or aboveground storage areas and then drives electricity-producing turbines when it is released. "There's some very good new technology on compressed air energy storage that can use existing gas turbine designs," said Steven Specker. "There's a plant in Alabama that has been operating for a number of years with compressed air energy storage. It works very well. [And] as wind power grows rapidly in certain parts of the United States and the world, we need a storage approach."

Today, intermittent sources of electricity, such as wind or solar power, can make electricity grids unstable.[1] To manage large amounts of renewable power generation, smart grids using advanced power electronics and electrical storage devices are needed to manage the transmission and distribution of electricity from where it is produced to where it is being used. Specifically, these grids must be able to "communicate" between utilities and electricity meters, enabling such advances as provision of power to the grid by plug-in hybrid automobiles and powering off of home appliances during peak load periods.

ENHANCED GEOTHERMAL ENERGY

Several speakers at the summit discussed sustainable sources of energy that are farther in the future than the expanded use of solar power and wind power. One is energy from engineered geothermal systems. As Dan Reicher pointed out, if a hole is drilled into the ground, it eventually will encounter rocks that are heated by Earth's interior. If several such holes are drilled near each other and the rock between them is fractured, water can be injected into one well and returned from the others much hotter, and steam from that water could turn a

[1]This depends on local grid characteristics, but in general grids are able to handle between 10 and 20 percent of intermittent renewables without requiring storage.

turbine to generate power (Figure 8.1). "It is a vast and ubiquitous base load resource, unlike solar and wind, which are obviously an intermittent resource," said Reicher. "If we can figure out how to exploit it, it could be developed in the large megawatt range."

There are no scientific showstoppers to exploiting this resource, said Reicher, but there are "great engineering challenges." No such system has been developed anywhere in the world; existing geothermal energy sources rely on heated water and steam that is near the surface of the earth. Furthermore, exploiting this resource requires that holes be drilled as deep as 10 kilometers into the earth and that rocks be fractured at great depths.

Can such deep wells be drilled? "The answer," said Reicher, "is that the oil industry does drill to those depths. And the oil industry does fracture rock at those depths. Ten kilometers is not an insignificant piece of work, but it is a distance the oil industry knows how to get to."

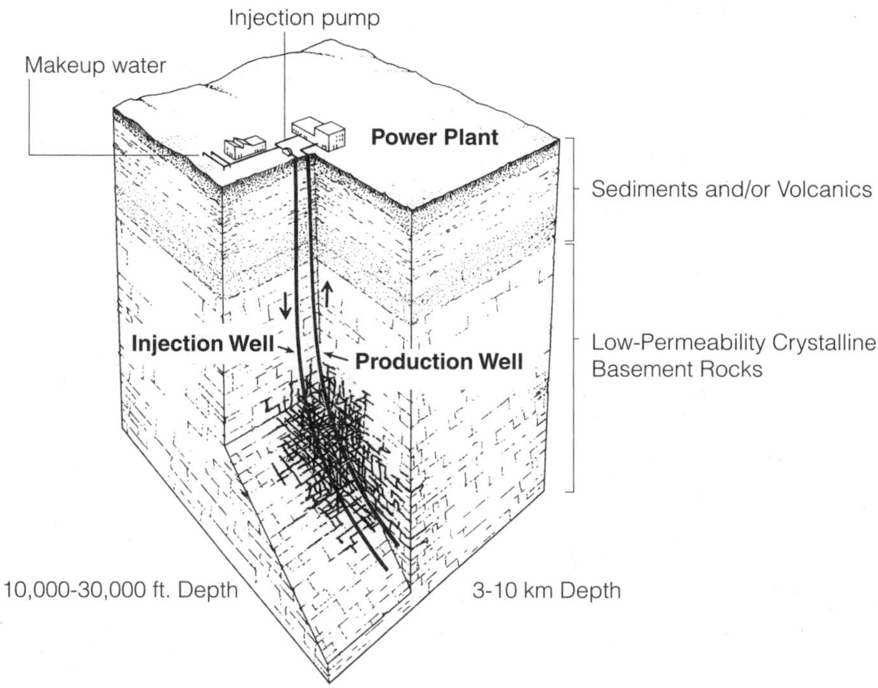

FIGURE 8.1 Enhanced geothermal systems could extract energy from Earth's interior. SOURCE: INL (2006).

Furthermore, an array of technological advances could cut costs, and the industry is working on these advances. For example, seismic technology can provide a good snapshot of fractured rock at depth to enable large areas underground to be connected, creating what is essentially a heat exchanger. Advanced drilling, control, and high-temperature technologies are all being investigated.

Today the Energy Department spends only $20 million per year to pursue this option. "This effort should be significantly expanded," said Reicher. "The Australians, who lead the world in this technology, are spending hundreds of millions of dollars. There are 30 companies in Australia working at this today, and we're playing catch up."

ADVANCED ENERGY R&D

Orbach also discussed ways of generating energy that will require what he called "transformational discoveries" in basic research. Over the past 5 years the Energy Department has conducted a series of workshops on basic research needs for a secure energy future, examining such topics as superconductivity, the hydrogen economy, solid-state lighting, advanced nuclear reactor designs, energy storage, and materials science. Using scientific and engineering research, Orbach asked, "What can we do to break out of the straightjacket in which we find ourselves?" For example, he mentioned studies of photosynthesis as a possible way to take advantage of techniques that living things have evolved to meet their energy needs. "It really comes down to how nature works," he said. "Plants are almost 100 percent efficient at room temperature. Is there any way for us to mimic what nature does so well?"

The technologies of this century will be rooted in the ability to direct and control matter at the molecular, atomic, and quantum levels, according to Orbach. Research challenges include the synthesis of new forms of matter with tailored properties, predictions of the properties of novel materials, and the fabrication of manmade nanoscale objects with capabilities that rival those of living things. Incremental changes will not be sufficient, Orbach said. Transformational discoveries and disruptive technologies will be essential.

Part III

Energy Uses

Maximizing energy efficiency and decreasing energy use will remain the lowest hanging fruit for the next several decades. It is something that we should do and we must do.

—Steven Chu

9

Transportation

The United States currently imports about two-thirds of the petroleum that it uses. This is about the same amount of petroleum that is burned in U.S. vehicles. There is an "absolute compelling need to do more in the area of energy efficiency and energy-enhancing technology development," said Reuben Jeffrey, and improving the efficiency of vehicles will be an essential part of that task.

PROSPECTS FOR IMPROVEMENT

The fuel economy of the new light-duty vehicle fleet (cars, light trucks, and sport utility vehicles) has declined over the past two decades, Paul Portney observed. "We haven't been making a lot of progress in making cars that reduce our dependence on oil overall or on imported oil." But Portney expressed several reasons for optimism. In recent years, fuel economy standards for the light-duty truck segment of the new vehicle fleet have gone up twice. More importantly, the Energy Independence and Security Act of 2007 will require automakers to achieve an average 35 miles per gallon for new cars and light-duty trucks by the year 2020, up from about 25 miles per gallon for new cars and light trucks today. That legislation was a major achievement, Portney said, given the political climate in the United States and the many challenges facing domestic car manufacturers.

Furthermore, considerable additional progress is possible. Rodney Nelson cited a recent study done by a Massachusetts Institute of Technology group which concluded that it is technically possible to double the fuel economy of

U.S. cars and light-duty trucks by 2035 (Cheah et al., 2007). "So the recent legislation went partway. It's technically possible to go farther," Nelson said. "It costs money, and the consumer may not necessarily jump up and down to buy one, but it's technically possible. And it's the biggest lever the United States has to decrease oil demand."

High gas prices are very painful to people who cannot afford them, Portney said. But high prices also have a positive effect by changing incentives, technologies, and tastes. Gasoline consumption, which had been rising for many years, has begun to decline. And a different mix of vehicles is appearing on U.S. roads. "Hybrids on the road will soon number in the millions rather than the tens or even hundreds of thousands," Portney said. "We've seen the introduction into the United States on a larger scale of cleaner, much better performing diesel engines, which is a very important development in improving fuel economy. And now we've begun to see what I think is perhaps the most promising of the so-called conventional technologies—plug-in hybrids, which I think will change the fuel economy picture in the United States significantly."

Another positive effect of high gas prices is that venture capitalists have become much more focused on clean energy and efficient vehicles. They are investing in new battery technologies, lighter materials, and alternative fuels, among other technologies. (The role of the private sector in the energy market is discussed in Part IV.)

VEHICLE TECHNOLOGIES

Even beyond the projections of the MIT group, great potential exists to improve the efficiency of vehicles and reduce U.S. dependence on imported petroleum, said Amory Lovins, who emphasized the energy savings that could be realized through a wide variety of technological innovations. Indeed, the key to energy efficiency, according to Lovins, is to realize that efficiency is profitable, not costly. "It is cheaper to save fuel than to buy fuel," Lovins said. "The climate debate should be about profits and competitive advantage. Once it is reframed in that way, any remaining resistance will melt faster than the glaciers. The biggest obstacle is the assumption that climate protection is costly. That is unexamined and clearly untrue, as many smart firms demonstrate [by] making billions of dollars substituting efficiency for fuel."

Lovins' Rocky Mountain Institute has estimated that the efficiency of cars, trucks, and planes could be tripled with investments that would pay for themselves in 2 years, 1 year, and 4 or 5 years, respectively, if decisions were made to pursue these efficiencies. First, vehicles could be made ultra-lightweight, more efficiently powered, and "slippery," so that they move through the air and along the road with less resistance (and often with better performance). For example, a diesel-hybrid carbon-fiber concept car from Opal can go 155 miles per hour and get 94 miles per gallon (though not at the same time). Surprisingly, the ultra-lightweight construction does not increase mass production costs, because the costlier materials are offset by simpler automaking and a power train that is three times smaller.

A major problem with a conventional car, Lovins said, is that seven-eighths of the energy it uses never gets to the wheels. It is consumed in the engine, the driveline, and accessories, as well as in idling. Half of the remaining eighth either heats the tires and road or heats the air through which the car passes. "Only the last 6 percent actually accelerates the car and then heats the brakes when you stop," Lovins said. Furthermore, only a twentieth of the mass in a car is the person driving it. The rest is the heavy steel car. So only 0.3 percent of the fuel burned by an automobile ends up moving the driver. "This is not very gratifying after 120 years of devoted engineering effort."

Because of this inefficiency, there is enormous leverage in making cars lighter, whether through light metals or advanced polymer composites. A group at the Rocky Mountain Institute completely redesigned a midsize car that can comfortably carry five adults and has 2 cubic meters of cargo space. It weighs less than half as much as the typical car today but still can protect its passengers when run into a sport utility vehicle twice its weight or into a wall at 35 miles an hour. It gets the equivalent of 114 miles per gallon on a fuel cell or 67 miles per gallon as a hybrid with a power train like a Prius, with "quite brisk performance," according to Lovins. The estimated sticker price is $2,500 higher than a conventional car, not because it is ultralight but because of its hybrid engine. Overall, the car is cheaper to manufacture because it has only 14 body parts, which are suspended from rings rather than built up from a tub (Figure 9.1). "This is like an airframe, not a horse and buggy," Lovins said. Most of the body parts, which are made from a single low-pressure die set, can be lifted by hand. A steel body would have 10 or 20 times more parts, and each part would have an average of four steel-stamping die sets. The parts precisely snap together for bonding, which means the usual jigs, robots, and welders are not needed, so that manufacturing plants are much less capital intensive. With color in the mold, even paint shops are unnecessary.

Lovins displayed a test piece for military helmets that was two-thirds carbon fiber and one-third carbon plastic, which was stronger than titanium. "Plastics have changed since *The Graduate*," he said. Such materials can provide aerospace performance at automotives costs. Cars made of such materials are half the weight and save half the fuel. Such materials also absorb 12 times more crash energy per pound than steel, with manufacturing costs about the same as for steel. Lovins acknowledged that composite materials are not the only possible solution; metals can offer some of the same advantages. "The market will sort out which ones win."

Lovins also described a concept car developed by Toyota (Figure 9.2). It has the interior volume of a Prius but uses half as much fuel. It weighs 400 kilograms (880 pounds) as a hybrid, about one-third the weight of a Prius, or

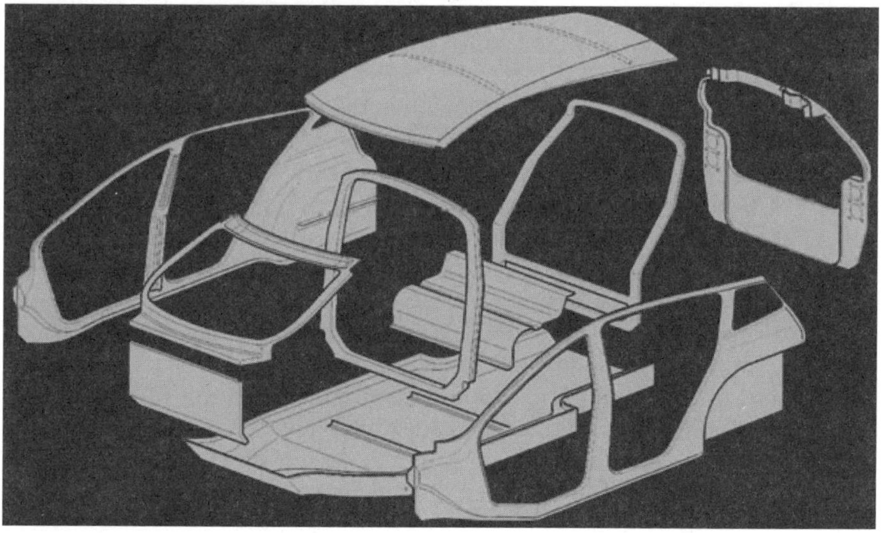

FIGURE 9.1 Lightweight cars based on composite materials could lead to radically simplified manufacturing. SOURCE: Fiberforge Corporation and Rocky Mountain Institute.

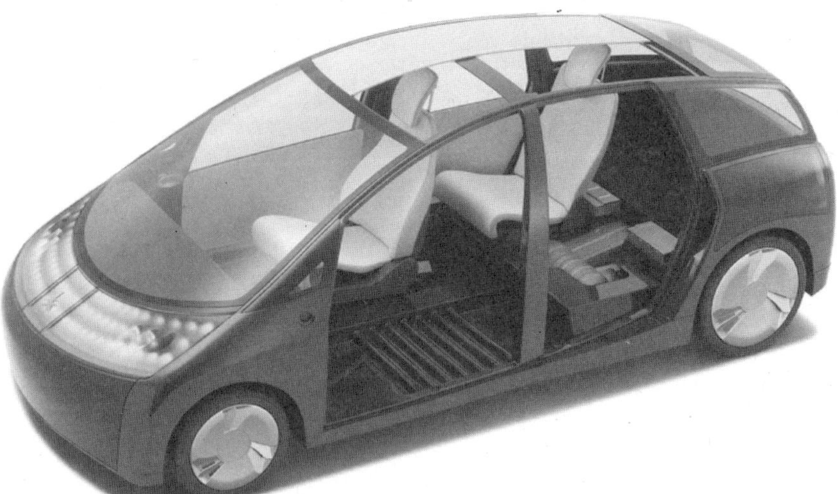

FIGURE 9.2 The 1/X concept car, designed by Toyota, that weighs one-third as much as a Prius, uses half as much energy, and yet has the same interior volume. SOURCE: Toyota Motor Corporation.

20 kilograms (44 pounds) more with batteries added to the car to make it a plug-in hybrid. "Coincidentally, [this] is exactly what I said in 1991—I think to an Academy group—that a good four-seat carbon car should weigh, to much hilarity from the industry."

Concept cars are often dismissed as bragging, Lovins said, but in this case the company Toray recently announced that it will build a ¥30 billion company to mass produce carbon-fiber autobody panels and other parts for companies like Toyota and Nissan. About the same time, Ford announced that it will reduce the weight of every platform it makes by 250 to 750 pounds starting with the 2012 model year. The following month, Nissan announced an average weight reduction of 15 percent by 2015. "Light-weighting is now the hottest strategic trend in the industry," Lovins commented.

These and other changes in the transportation sector could have a substantial impact on oil imports, Lovins said. Combined with the use of biofuels, the substitution of natural gas for oil, and the generation of hydrogen from renewable energy sources like wind, the projected consumption of oil could be cut from 28 million barrels per day in 2025 to 16 million barrels per day using technologies that are already available, he said (Figure 9.3). Cumulative carbon emissions would go down by more than a trillion tons, and the amount spent to buy petroleum would be reduced by tens of trillions of dollars. Further efficiency gains, greater use of biofuels or natural gas, or the use of hydrogen from natural gas or renewable energy sources could eliminate the need to import any petroleum. (Pathways to a sustainable energy future are described in Part IV.)

Even major industries have changed quickly in the past, Lovins pointed out. The auto industry in the 1920s took just 6 years to switch from wood to

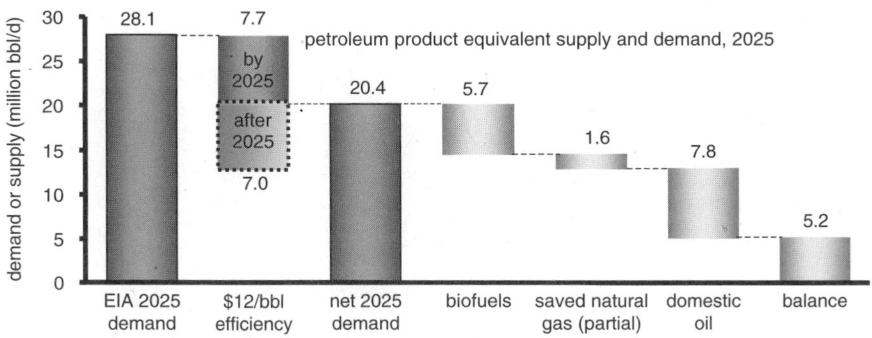

FIGURE 9.3 Greater efficiency (the second bar in the above graph), combined with greater use of biofuels and natural gas, could reduce or eliminate U.S. reliance on imported petroleum. SOURCE: Rocky Mountain Institute.

steel auto bodies. At the start of World War II, it took just 6 months for the industry to switch from making cars to the tanks and planes that won the war. Boeing instituted a radically new airframe in the 787 in just 5 years. A small team at General Motors took a battery-powered car from launch to the street in 3 years. Technology diffusion normally takes longer—12 to 15 years for a new technology to go from 10 to 90 percent adoption—but the right policies and innovative business strategies can make adoption much quicker.

"Can the U.S. auto industry turn itself around through advanced designs that integrate light materials and new forms of manufacturing?" Lovins asked. Boeing's experience is evidence that it can. A decade ago, Boeing was "in as deep a crisis as Detroit is now." No great innovations were in the production pipeline, Airbus was pulling ahead of Boeing, and some people were questioning the company's future.

Boeing's response was to begin work on the 787 Dreamliner. It used 20 percent less fuel but cost the same. Half of its mass consisted of carbon composites, up from 9 percent in other planes, with bigger windows, a higher-pressure cabin, and other advantages to passengers. "It has had the fastest order takeoff of any jet in history," Lovins said. "It is sold out well into 2015, and its suite of innovations is being rolled into every plane Boeing makes."

Lovins said that the energy use of six sectors has to change, with at least three probably already past the tipping point. In aviation, Boeing was successful and is now ahead of Airbus. In heavy trucks, Wal-Mart has been leading the way, with much more efficient trucks being introduced into the market where users can buy them. In the military, the Pentagon has emerged as a federal leader in reducing dependence on petroleum. In fuels, there is strong investor interest and industrial activity. And in finance, there has been growing interest in the energy sector, with the clean energy sector getting $117 billion of private risk capital last year.

The sixth sector—cars and light trucks—is the toughest sector to change, but progress is being made. The head of Boeing's commercial airplane division recently moved to Ford to bring ideas from the aerospace industry to the automotive industry. The unions and automobile dealers are keen for innovation, Lovins said, "to save the industry as a tsunami of creative destruction sweeps over them." The Rocky Mountain Institute has been doing projects with the auto industry. "The level of competition now is producing unthinkable change, and it's going to change the automakers' managers, or their minds, whichever comes first."

Governmental policies can hasten these changes. For example, consumers tend to have a very high implicit discount rate, Lovins said, in that they do not consider the fuel savings beyond the first year or two. One way to build the discount rate into the price of a vehicle is through what are called feebates. Consumers pay a fee for buying less efficient vehicles, and the proceeds of that fee are used to offer rebates on more efficient vehicles. "That turns out to be extremely powerful and is more profitable for the automakers," Lovins said, "because to move their offerings from the fee zone into the rebate zone, they add technology content that has a higher [profit] margin than the rest of the vehicle." The District of Columbia has such a system. "It's not a feebate, but it walks and quacks like one," Lovins said. The sales tax for very efficient cars goes to zero, while the sales tax on heavy and inefficient cars, which wear out the streets faster, is higher. It's an example that "illustrates the opportunities for state-level leadership and experimentation in innovative public policy instruments." Also, the industry is increasingly interested in feebates as a way of getting more efficient cars on the road faster than with gasoline taxes or standards.

As a final example of what is possible, Lovins mentioned the fifth for-profit company that has been spun off from the Rocky Mountain Institute—a company designed to bring a lightweight plug-in hybrid vehicle to market. He said that if half the cars in the country were plug-in hybrids, wind plants could supply them with power, producing as much or more power than all of the nation's coal plants do now.

INITIATIVES UNDER WAY

Steven Chu also insisted that new technologies can make a big difference in the transportation sector. For example, in normal use the charge of a battery in a Prius ranges between 40 and 60 percent of its maximum. For a plug-in hybrid or all-electric vehicle, a battery is needed that can discharge 80 or 90 percent of its capacity.

For almost a decade, the auto companies and the Department of Energy collaborated on a solid lithium metal battery that had a much higher energy density and could be discharged and recharged more effectively. However, problems with the battery caused the program to be abandoned. Recently, researchers at Lawrence Berkeley National Laboratory have combined a polymer used in the former program with a new compound that overcomes many of the battery's problems. "After 1,000 deep discharge cycles, 90 percent discharge, [there is] no sign of wear," Chu said. Extrapolating from that result, such a battery could last for perhaps 10,000 discharge cycles before starting to show signs of wear at 15 years.

Dan Reicher also mentioned an effort under way at Google known as RechargeIt. A fleet of hybrid vehicles are being driven by Google employees, with data on their experiences being posted on a publicly available website. The vehicles plug into one of the largest photovoltaic systems in the United States, which covers a large parking structure and many roofs of the company's buildings.

Finally, Ray Orbach mentioned the potential still to be derived from conventional technologies. "Don't bet against the internal combustion engine," he said. "[Engineers] have worked remarkable things with that engine. . . . People are coming up with electric drives, but what do we use? Hybrids. I think the internal combustion engine is going to be around for a while."

HYDROGEN AS AN ENERGY SOURCE

Hydrogen has many advantages as a fuel for vehicles and as an energy source for buildings, industries, and other energy users, according to Michael Ramage. Its use could reduce petroleum consumption in the United States, making the nation less reliant on foreign sources of oil. It also can be generated from sustainable energy sources, resulting in substantial reductions in greenhouse gas emissions.

However, hydrogen also has many hurdles it must overcome before it assumes a major role in any nation's energy portfolio, Ramage added. Hydrogen must be generated from other energy sources, which can result in greenhouse gas emissions, reliance on uncertain suppliers, or other problems.[1] The substantial use of hydrogen would require a massive change in the energy infrastructure of the country. Today, generating hydrogen remains more costly than other sources of energy, and safety issues need to be resolved. Finally, "hurdles to the hydrogen economy are much more than technical," said Ramage. "They are political and social."

An emphasis on specific problems confronting hydrogen has generated concern and controversy, Ramage noted. Only by viewing hydrogen energy as an integrated system, from production to distribution to use, can its potential as an energy source be assessed.

Hydrogen can come from many sources, including fossil fuels, nuclear power, solar power, wind power, or biomass (Figure 9.4). Researchers also are looking at more advanced sources of hydrogen, such as electrolysis of water by genetically engineered microorganisms. Each of these sources has its own advantages and disadvantages. For example, production from fossil fuels releases greenhouse gases, which must either be captured and sequestered or released into the atmosphere. Also, a major source of hydrogen at least initially will be natural gas. But domestic supplies of natural gas are limited, which means that the United States would have to import more natural gas from other countries.

Hydrogen production plants can exist at different scales (Figure 9.5). A large central plant could produce on the order of 1 million kilograms of hydrogen per day—enough to fuel about 2 million cars. (A kilogram of hydrogen has about the same energy content as a gallon of gasoline.) Such a plant is about

---

[1]For a full discussion of hydrogen production, see Chapter 6 in NRC and NAE (2004). For more information on the use of hydrogen as an energy source, see NRC (2008a,b).

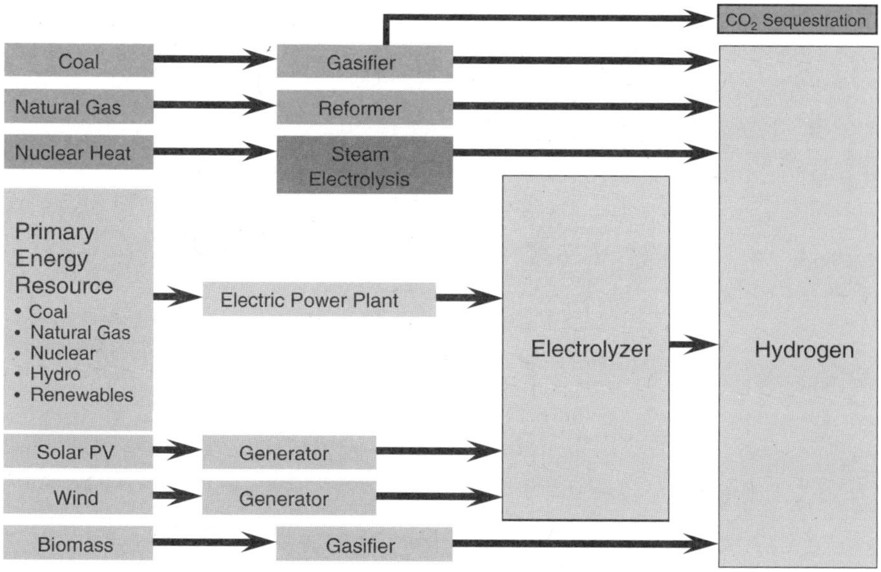

FIGURE 9.4 Hydrogen can be produced from many sources. SOURCE: Courtesy of Martin Offutt, IIASA, based on data from L. Burns, General Motors.

one-tenth the size of a large U.S. refinery, Ramage said. "For reference, Los Angeles has 10 million cars, and so would need five plants of this size."

A midsize plant could produce enough hydrogen for about 43,000 cars. Such a plant would be a particularly good size for biomass production, Ramage said.

Hydrogen production also can be carried out in many small plants, including filling stations for vehicles. The hydrogen could be made at the site through hydrolysis, or it could be generated directly from natural gas. Today the cost of the amount of hydrogen equivalent to a gallon of gasoline made at filling stations from natural gas is about $3.00, assuming a price for natural gas of $6.00 per cubic foot. Such a facility could serve somewhere around 1,000 cars a day.

Hydrogen produced from coal would be competitive with gasoline today. But the critical issue is whether the carbon dioxide from the production process would be captured and sequestered. Producing hydrogen from biomass has been getting cheaper, but much more research needs to be done on the use of biomass as an energy source. Considerable research also is needed on the supply system for hydrogen.

Hydrogen often will be used as an energy source to power fuel cells that will produce electricity. The performance of fuel cells is therefore a critical issue

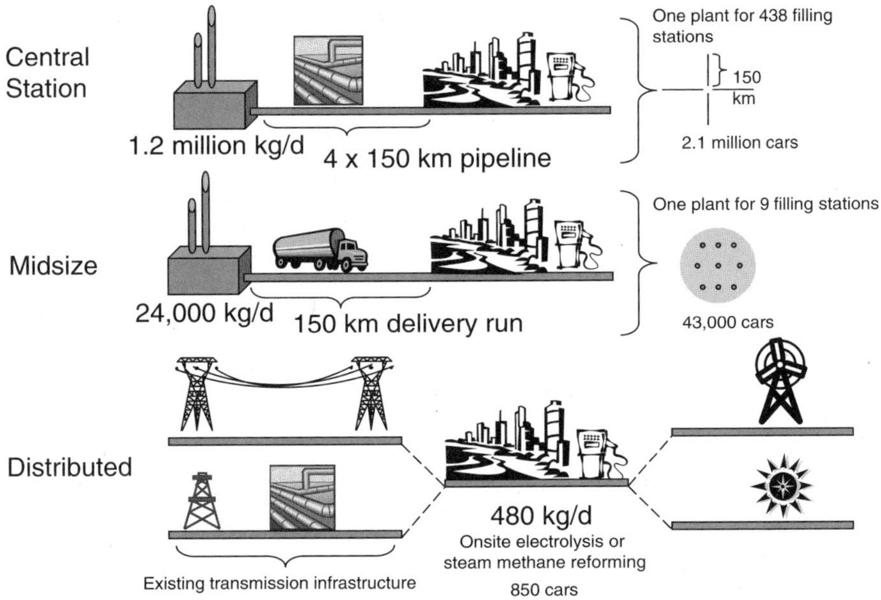

FIGURE 9.5 Hydrogen production plants can be built on different scales. Courtesy of Martin Offutt, IIASA.

both during the transition to the widespread use of hydrogen and in the longer term. The cost of power from a fuel cell has fallen from $1,000 per kilowatt in the early 1990s to $300 per kilowatt in the year 2000. For a mass-produced fleet of 500,000 vehicles, the cost would be about $100 per kilowatt today. A feasible but ambitious target would be $30 per kilowatt.

The durability of fuel cells also has improved, from 1,000 hours in 2004 to 2,000 hours in 2007. A good target for durability is 5,000 hours, Ramage said.

A major challenge for vehicles is to get enough hydrogen onboard the vehicle for it to have a range of about 300 miles, comparable to current vehicles. The best solution appears to be absorbing hydrogen in solid materials, but "this is a major technical challenge," according to Ramage. At present, automakers are prepared to take vehicles to market that use hydrogen compressed to 10,000 pounds per square inch (psi), and there are already a growing number of demonstration vehicles on the road that have ranges on the order of 300 miles.

In the optimally plausible case, which assumes that technical targets are met, that policies are established to support infrastructure change, and that consumers buy the vehicles, hydrogen fuel cell vehicles could make significant inroads into current vehicle markets, Ramage said. If they have a penetration rate comparable to that of hybrid vehicles and then front wheel drive vehicles, there would be a substantial transition toward hydrogen vehicles in the 2025 to

2040 time period (Figure 9.6). Such a trend would require a substantial increase in hydrogen production during this period, which would lead to a corresponding reduction in petroleum imports. "Obviously, nobody would assume that hydrogen is going to be our only source of energy," Ramage said. But such a scenario "shows the kind of impact it could have."

The transition to a new energy regime is discussed in Part IV of this summary, but several points are particular to hydrogen. If hydrogen vehicles become widely available in 2015 and are self-sustaining in 2025, which means that they are cost-effective and people are buying them without government subsidies, the best method of hydrogen distribution during that period will be fueling sites using natural gas. "Natural gas is an energy security issue," Ramage said, but "you can get a lot of vehicles on the road in the early years with small refining units at the filling station, and most of that technology is in place."

The technology to produce hydrogen from biomass and from coal with sequestration could be developed in the 2020-2025 timeframe, Ramage said. Other large-scale technologies, such as nuclear, solar, and other renewables, then could be developed over a longer timeframe. Under an optimally plausible scenario, U.S. petroleum consumption could be reduced 40 percent by 2035. Energy security would be enhanced, light-duty vehicles would emit half as much carbon dioxide, and 20 percent of the overall U.S. energy used could come from hydrogen. "Equally important," said Ramage, "it's a pathway to a sustainable energy future."

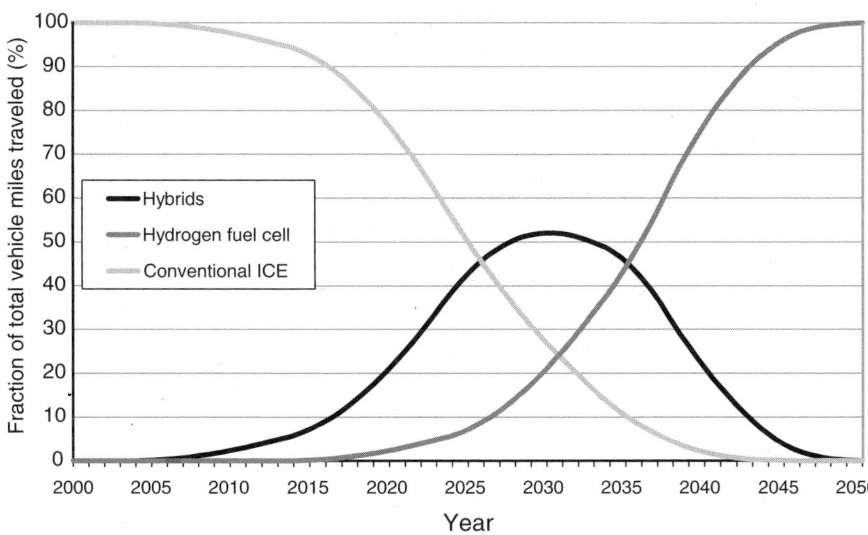

FIGURE 9.6 Under an optimistic scenario, fuel cells could account for virtually all vehicle miles traveled in the United States by the middle of the century. ICE, internal combustion engine. SOURCE: Based on Figure 3.1 in NRC and NAE (2004).

"None of us can really imagine what the energy future is going to look like," said Ramage. But the transition is sure to require public-private partnering. It also will require public incentives to buy hydrogen-powered vehicles, such as are in place for hybrid vehicles. And the incentives need to be large enough and sustained enough to get through the transition to widespread use of hydrogen.

A robust, ongoing program of research and development by both the public and private sectors could make progress on the hurdles hydrogen faces, Ramage said. To name just a few issues: Vehicle and fuel combinations need to be evaluated for their impact on enhancing energy security and reducing carbon dioxide emissions. Details of the transition to hydrogen need to be studied. And strategies to accelerate the innovation process for the needed technologies need to be studied.

10

Buildings and Industry

Beyond transportation, tremendous amounts of energy can be saved in homes, businesses, and industry through more efficient technologies and through changes in behaviors. "The world's energy needs and our environmental responsibilities converge toward a common solution," said Reuben Jeffrey. "Conservation must remain an important part of that solution."

Samuel Bodman made a similar point, "We all must actively promote enhanced energy efficiency wherever we can—in our homes, our vehicles, our offices, and across all industries. Because the truth is, the largest source of immediately available, 'new' energy is the energy we waste every day." The needed steps are straightforward, said Bodman: insulating homes and other buildings, choosing energy-efficient appliances and compact fluorescent bulbs, considering a fuel-efficient vehicle or taking public transportation, and participating in an energy assessment program. "Collectively, these actions have an impact in precisely the right direction, taking some immediate pressure off demand."

The changes going on in U.S. society today are more than a reaction to higher gas or home-heating costs, Bodman said. "I believe we are seeing a growing commitment to not just affordable energy but clean energy as well."

Speakers at the summit discussed just a few of the many efficiency improvements possible through technological change. But these examples demonstrate a point that extends much more broadly—efficiency improvements often pay for themselves over time, making reduction of greenhouse gas emissions through energy efficiency a win-win proposition.

CONSERVATION IN CALIFORNIA

California has been leading the way in conserving electricity, Steven Chu pointed out. In 1973, electricity use in the United States was about 8,000 kilowatt-hours per person for the country as a whole and about 6,500 kilowatt-hours for California. By 2006, electricity use per person in the United States averaged 12,000 kilowatt-hours, while in California electricity use was about 7,000 kilowatt-hours. Meanwhile, the real gross domestic product of California grew by a factor of two. It's a "myth that if you flatten the use of energy you will kill your economy," Chu said.

One of the most important advances in California was the introduction of energy efficiency standards. For example, the efficiency of refrigerators improved by more than a factor of four from 1975 to 2005 even as the size of refrigerators grew from 18 cubic feet to 22 cubic feet. Meanwhile, the inflation-adjusted cost of refrigerators went down by a factor of two. The amount of energy saved just from refrigerators is equivalent to more than two-thirds of all the hydroelectric power generated in the United States (Figure 10.1). Major efficiency gains also are possible with air conditioners and gas furnaces. "This should be done throughout the whole sector," Chu said.

Large amounts of energy also can be saved through building codes. California has a temperate climate, which makes heating and cooling buildings less expensive. Yet California codes call for extensive building insulation, which

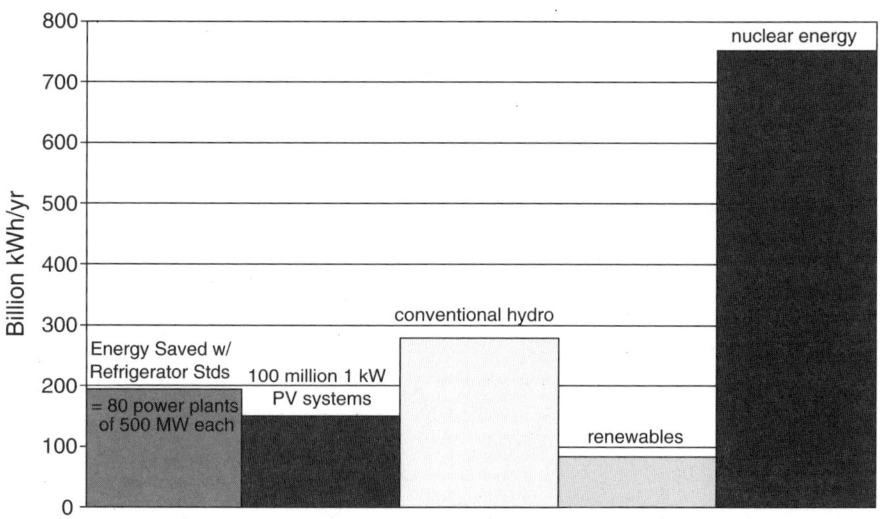

FIGURE 10.1 The amount of energy saved by efficiency standards for refrigerators compared to several other sources of supply in the United States. PV, photovoltaic. SOURCE: Courtesy of Arthur H. Rosenfeld, California Energy Commission.

has helped to conserve energy in the state. "There are so many opportunities to capture energy efficiency in buildings," Chu said. "We're talking factors of three or more in new buildings that would pay for themselves in 5 years."

An important policy innovation that has spurred conservation in the state has been to separate profits from energy sales. Traditionally, utility companies have made more money by selling more energy, creating an incentive to spur energy consumption. State policy in California consciously decoupled that connection. Instead, if utilities introduced more efficiency into the use of energy, they were able to charge higher rates. "There were inducements for the utility companies to actually make investments and save more energy," Chu said. This has had a profound effect by making utilities outspoken advocates for energy efficiency in the state. For many years, California was the only state to have taken such an action, but it now is being implemented by states elsewhere.

Enlightened policies also can affect individual perceptions of discount periods. Most people are not impressed by an investment that will pay for itself in 20 years. They tend to like to see a repayment within a year and a half. But if they can be repaid in 5 to 6 years, inducements and policies that promote awareness of such payoffs can create "a big change," said Chu. For example, a recent report by McKinsey & Company (2007) estimated that $1,000 worth of additional insulation and labor could pay for itself within 1½ to 2 years. Yet the American Home Builders Association is lobbying very strongly against efficiency targets, said Chu, contending that "American homeowners . . . don't want to pay for [improvements]." The solution, according to Chu, is "to write your Congress people."

Finally, Chu lauded the Top Runner program in Japan, which he called "an Energy Star labeling program on steroids." The program identifies the most efficient product in a variety of categories and then uses the performance of this "top runner" model to set a target for all manufacturers to achieve within the next 4 to 8 years. "You don't really need an elaborate appliance standard bureaucracy," Chu said. You let "industry bootstrap itself. It's a target, not a mandatory regulation, yet it seems to be very effective."

Chu asked why programs like California's utility company decoupling or Japan's Top Runner have not been widely replicated around the world. Partly it is because people have not heard about the programs. In a 2007 report entitled *Lighting the Way: Toward a Sustainable Energy Future* (IAC, 2007), a committee cochaired by Chu and José Goldemberg recommended that a small international committee of experts be used to identify policies that have worked. Such policies are like the rudder of a ship that governments can use to produce enormous course changes over time.

SAVING MONEY BY SAVING ENERGY

The Rocky Mountain Institute also has examined a large number of ways to improve efficiency in homes and businesses. According to Lovins, instituting economical ways to save electricity could save about three-quarters of the electricity consumed in the United States at an average cost of about 1 cent per kilowatt-hour, which is cheaper than the cost of operating a power plant. Other studies have arrived at comparable findings, and "the efficiency potential keeps getting bigger and cheaper because the technology improves faster than we use it," Lovins said. "It's as if the low-hanging fruit is mushing up around our ankles and spilling in over the tops of our waders while the innovation tree pelts our head with more fruit."

Lovins used his house and Rocky Mountain Institute headquarters in Snowmass, Colorado, as an example (Figure 10.2). At an elevation of 7,100 feet in the Rockies—where frost is possible any day of the year, winters can be continually cloudy, and lows can reach minus 47 degree Fahrenheit—Lovins has harvested 28 banana crops in the central atrium. Yet the house does not have a furnace. It is extremely well insulated and efficiently designed, which together cost less than a heating system would have cost. The building saves 99 percent of space and water heating energy and 90 percent of home electricity, and when the house was built in 1983 the additional construction costs paid for themselves in 10 months.

Other projects have demonstrated similar results in extremely hot climates, Lovins said. Homes in such climates can be built with no air conditioner and remain quite comfortable in temperatures up to 115 degrees Fahrenheit. Even in humid Bangkok, homes can use 90 percent less air conditioning and still offer better comfort and no extra construction costs. "These examples span the range of Earth's climates and tell a common story," Lovins said. "If you optimize the house as a system, . . . you get big, cheap savings." The object is to use technologies and smart design to tunnel through the cost barrier of diminishing returns and rising marginal costs. "If I add enough [insulation], I get rid of the furnace, ducts, vents, pipes, wires, controls, and fuel supply arrangements. It's 99 percent cheaper than if I had set out to save little or nothing."

Rocky Mountain Institute has demonstrated these steps in $30 billion worth of industrial projects in 29 sectors. Sometimes the changes are as simple as designing production plants to use fat, short, and straight pipes rather than thin, long, and crooked pipes. "It works better and costs less," Lovins said.

TRANSFERRING TECHNOLOGIES TO THE DEVELOPING WORLD

Much of the growth in energy consumption over the next two decades will be in the developing world. Therefore, energy-efficient technologies need to be transferred from the developed world to the developing world, said Rodney Nelson. A major recommendation from the National Petroleum Council's study of world oil and gas supplies is that energy-efficient technologies need to be implemented outside the developed world (NPC, 2007).

FIGURE 10.2 *Top:* Amory Lovins' house (and the original headquarters of the Rocky Mountain Institute) in chilly Snowmass, Colorado (*left*), uses about 1 percent the normal space- and water-heating energy and 10 percent the normal electricity, with a 10-month payback in 1983. It has produced 28 indoor banana crops (*right*) with no furnace. *Middle:* A Davis, California, tract house, designed by Davis Energy Group to use about a tenth the normal U.S. amount of energy, is comfortable with no air conditioner at up to 115°F (45°C) and, if built in quantity, would cost about $1,800 less than normal to construct and $1,600 less over time to maintain. *Bottom:* This 350-square-meter Bangkok house, at normal construction cost, provides superior comfort with one-tenth the normal air-conditioning energy. SOURCE: *Top:* Courtesy of Rocky Mountain Institute. *Bottom:* Courtesy of Professor Suntoorn Boonyatikarn, Chulalongkorn University.

Sometimes these technologies can be very straightforward yet have a major impact, Chu observed. At the University of California, Berkeley, and Lawrence Berkeley National Laboratory, a team led by Ashok Gadgil developed a cook-stove that is four times more efficient than the three-stone stove traditionally used in Darfur (Figure 10.3). Each stove annually avoids the emission of 2 tons of carbon dioxide per year (a typical car emits 4 tons of carbon dioxide per year). It also produces much less indoor air pollution, which is responsible for the deaths of more than a million people worldwide each year. The cost is less than $20, including a modest profit to the local manufacturer. "The energy problem can be greatly advanced by pretty low-tech stuff," said Chu. "In the poorest part of the developing world, quite modest things can have a profound impact."

FIGURE 10.3 A cookstove designed in Berkeley, California, and manufactured in the Sudan avoids 10 tons of carbon dioxide emissions over its 5-year life. SOURCE: Roy Kaltschmidt, Lawrence Berkeley National Laboratory.

Part IV

Meeting the Challenge

Are we up to the challenge? There is no reason to be particularly optimistic, but time will tell.

—James Schlesinger

11

Pathways to a Sustainable Future

Achieving an energy regime that meets human demands while protecting the global environment will require changing the relationship between energy use and economic activity. As several speakers at the summit pointed out, these two measures are correlated (Figure 11.1). However, the correlation is not invariant.

From 1977 to 1985, the U.S. economy grew 27 percent while the nation's use of oil fell 17 percent. Oil imports fell by half, and imports from the Persian Gulf dropped by 87 percent. "It broke OPEC's pricing power for a decade, because we customers, especially in America, . . . found that we could save oil faster than OPEC could conveniently sell less oil," said Amory Lovins.

As Lovins pointed out, economic theorists have assumed that energy intensity in the world will fall by about 1 percent a year because of increasing efficiency. "If we could make that about 2 percent a year, it would stabilize carbon emissions with economic projections. If we could make that more like 3 percent per year, carbon emissions would fall and stabilize the climate fairly quickly."

Reductions in energy intensity of 3 percent a year may seem high, but they are not uncommon, Lovins said. The United States has cut its energy intensity by that much or more in many recent years, including 4 percent in 2006. California's energy intensity typically has dropped a percentage point faster than the U.S. average. China cut its intensity by more than 5 percent a year for a quarter of a century, although it recently "came off the rails" as it began using more energy-intensive basic materials. But if China were to make energy intensity a priority, as it is now beginning to do, the country could have 20 times the gross domestic product that it does today while emitting no

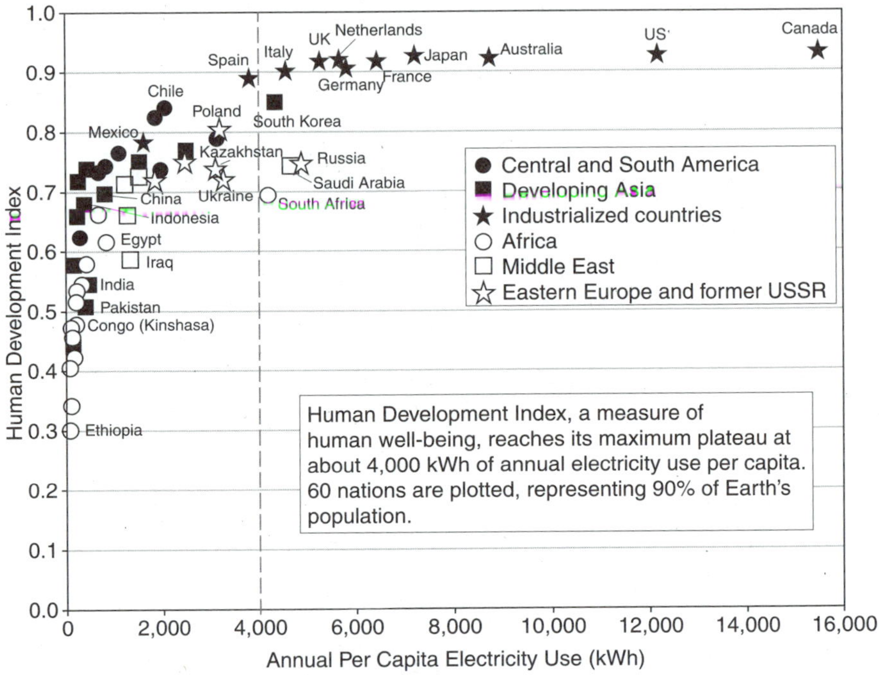

FIGURE 11.1 Annual per capita electricity use rises with the human development index to a maximum at about 4,000 kilowatt-hours. SOURCE: Adapted from Pasternak (2000).

more carbon, according to Lovins. Many companies have been cutting energy intensity—and in some cases absolute emission levels—by 6 to 9 percent a year. "They all make money on it," Lovins said. Even Japan, which has less than half the energy intensity of the United States, is finding ways in official studies to triple energy productivity

To solve the energy problem, the United States must increase its energy efficiency four- to fivefold, while the developing world grows in such a way that its energy intensity does not increase dramatically, said Steven Chu (Figure 11.2). "The real question is whether the developing countries will follow in the footsteps of the United States, Australia, and Canada," said Chu. Or will they "leapfrog past the mistakes of the developed world"? The developed world has an obligation to lead the way and to help other nations follow, Chu said. "It is not our birthright to say that we should enjoy a high standard of living and the developing countries should not."

Several speakers pointed out that stabilizing the amount of carbon dioxide in the atmosphere will require that carbon emissions be cut to a very low

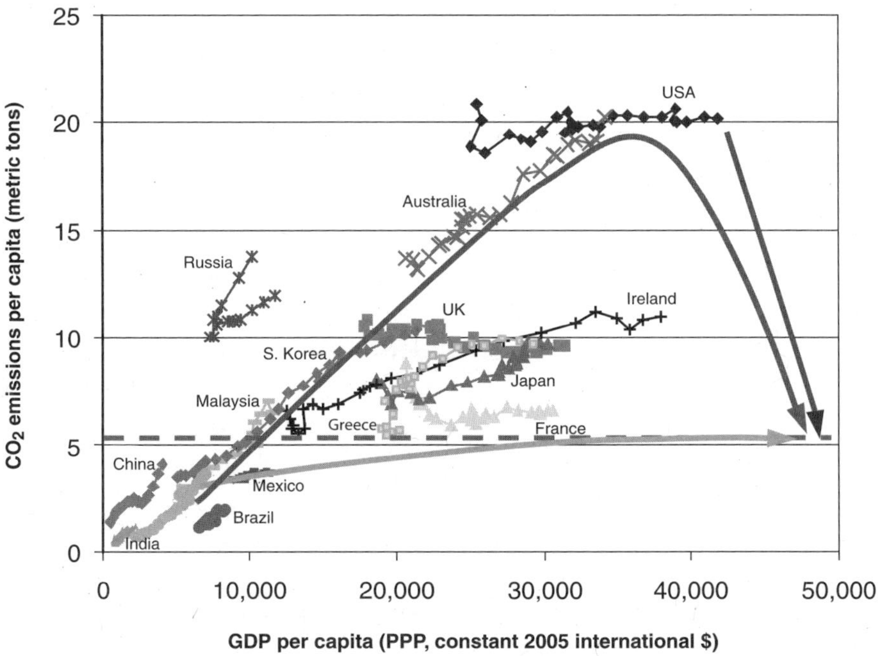

FIGURE 11.2 As the per capita gross domestic product of the developing countries increases, carbon dioxide emissions can either rise to the level of the most energy-intensive developed countries (upper curve) or remain at the level (lower curve and dashed straight line) that the developed world needs to reach to avoid dangerous climate change. PPP, purchasing power parity. SOURCE: Based on EIA and UN data plotted by members of the Office of the Chief Scientist, BP plc. GDP per capita data from the World Bank World Development Indicators 2008 database.

level—or eliminated entirely—in the United States and many other countries. "Zero [emissions] is the answer," said Robert Marlay. "Zero is a very inspiring technological goal, which has permeated all the thinking in the R&D agencies. This is what we need to imagine is possible. This is what we need to craft our vision and our programs to do. This is what we are going after."

As John Holdren said, "If you look at how long carbon dioxide stays in the atmosphere, we're going to have to be very near zero by the end of this century or shortly thereafter if we want the impacts of climate change to be manageable. And we're not going to avoid all of the impacts. I often say that in the climate challenge, we have only three choices—mitigation, adaptation, and suffering—and we're already doing some of each. What's up for grabs is the mix. If we want the suffering to be minimized, we're going to have to do a whole lot of mitigation and a whole lot of adaptation."

Several speakers at the summit described plans that would substantially reduce U.S. emissions of carbon dioxide. This chapter describes two of those plans. Steven Specker presented an analysis done by the Electric Power Research Institute (EPRI) that would reduce carbon dioxide emissions to levels below those for 1990 by the year 2030. Jon Creyts and Ken Ostrowski summarized a McKinsey & Company analysis (2007) that looked at more than 200 options for reducing carbon dioxide emissions. Although neither plan would reduce carbon emissions to anywhere near zero, both would "bend the curve" of U.S. emissions so that they begin to decline rather than continuing to increase.

ELECTRICITY TECHNOLOGY IN A CARBON-CONSTRAINED FUTURE

In plotting the future of electricity technologies given future constraints on carbon dioxide emissions, EPRI set out to answer three questions:

1. What is the technical potential for reducing U.S. electric sector carbon dioxide emissions?
2. What are the economic impacts of different technology strategies for reducing U.S. electric sector carbon dioxide emissions?
3. What are the key technology challenges for reducing electric sector carbon dioxide emissions?

The EPRI analysis focused on the period between now and 2030, since that is the period when technologies will have to be deployed to bend the curve of growing carbon dioxide emissions, Specker said.

Using projections from the EIA of carbon dioxide emissions over that period—which were recently modified to reflect the impact of the 2007 energy legislation—the EPRI study looked at the potential of seven technology areas to reduce emissions (Figure 11.3). The first area is efficiency. EPRI set a target of 0.75 percent growth for consumption in the electricity sector until 2030. That target is "aggressive but doable," said Specker. "If we can do better, that'll be fantastic, but we think that's a significant technical challenge." The best thing about efficiency improvements is that they can be started immediately. "You don't have to pour concrete. You don't have to build . . . new plants. There's a lot we can do with efficiency right now."

The second area EPRI considered is renewable sources of energy. The EIA has forecast that 60 gigawatts of such power would be available by 2030. The technology challenge set by EPRI is for 100 gigawatts. Together, efficiency improvements and additional sources of renewable energy "get pretty close, at least for a while, to flattening out carbon dioxide emissions in the electricity sector if we can achieve these targets."

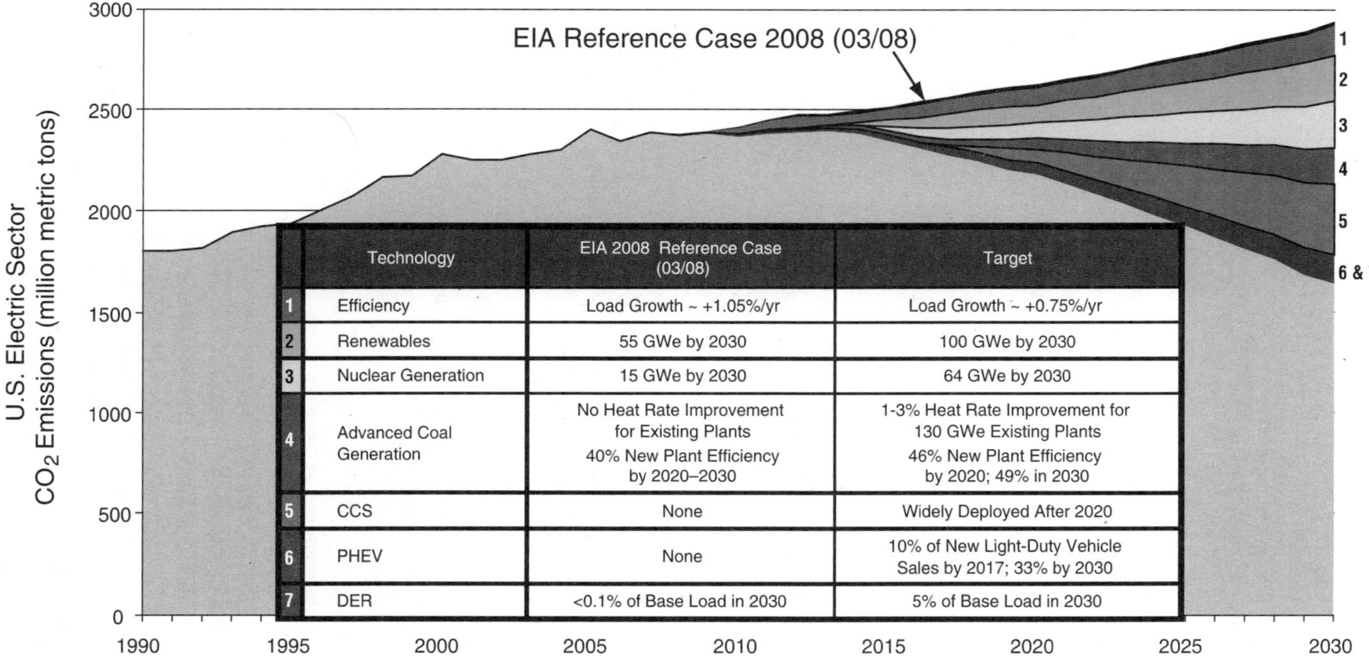

	Technology	EIA 2008 Reference Case (03/08)	Target
1	Efficiency	Load Growth ~ +1.05%/yr	Load Growth ~ +0.75%/yr
2	Renewables	55 GWe by 2030	100 GWe by 2030
3	Nuclear Generation	15 GWe by 2030	64 GWe by 2030
4	Advanced Coal Generation	No Heat Rate Improvement for Existing Plants 40% New Plant Efficiency by 2020–2030	1-3% Heat Rate Improvement for 130 GWe Existing Plants 46% New Plant Efficiency by 2020; 49% in 2030
5	CCS	None	Widely Deployed After 2020
6	PHEV	None	10% of New Light-Duty Vehicle Sales by 2017; 33% by 2030
7	DER	<0.1% of Base Load in 2030	5% of Base Load in 2030

FIGURE 11.3 Emissions of carbon dioxide by the U.S. electric generation sector could drop below 1990 levels by 2030 through the use of seven categories of technologies. NOTE: CCS, carbon capture and sequestration; PHEV, plug-in hybrid electric vehicle; DER, distributed energy resources. SOURCE: Energy Technology Assessment Center of the Electric Power Research Institute.

The third technology challenge is greater use of nuclear energy. Compared with the EIA forecast of 20 gigawatts of new nuclear capacity by 2030, EPRI has set a target of 64 gigawatts of new nuclear power by then. The first new advanced light-water reactors would come on line in 2016. Creating 64 gigawatts of new capacity would require 40 to 45 new advanced light-water reactors by 2030. When new nuclear capacity is added to efficiency and renewables, the curve of carbon dioxide emission starts to bend downward.

Advanced coal generation without carbon capture and sequestration is the fourth area. Two opportunities exist in this area. About half of the existing coal plants in the United States have the potential for efficiency improvements of 1 to 3 percent. That's the "quickest, easiest way to get carbon dioxide reductions in the existing installed base," Specker said. The second opportunity is to improve the technology of plants through higher temperatures and pressures to get efficiencies as high as 49 percent by 2030. This goal poses "lots of materials challenges," said Specker, but it is an important component of the overall plan.

The EIA reference case does not assume any carbon capture and sequestration because it is based on existing laws and regulations without a price on carbon. EPRI has set a goal of wide-scale deployment of advanced coal with carbon capture and sequestration by 2020—its fifth technological focus—that would require all new coal plants coming on line after 2020 to have up to 90 percent carbon capture and storage. This is "a very daunting technology challenge," said Specker, "but we think [it is] absolutely essential."

The sixth area is the widespread use of plug-in hybrid electric vehicles, an area in which EPRI has focused considerable attention in recent years. And the seventh and final area is the use of distributed energy resources, mostly solar photovoltaic energy, which could expand significantly in the latter part of the period EPRI considered.

With these areas of emphasis, carbon dioxide emissions can be reduced below 1990 levels by 2030, and the curve of increasing emissions can start to bend in the 2012 to 2015 time period, according to the EPRI analysis. "It's all about efficiency and renewables in these early years," Specker said. "But that's not going to be enough to do what we need to do long-term."

The EPRI analysis took a second approach to considering carbon dioxide emissions. It assumed that emissions would be limited in the future and asked how electricity production would have to change given those limits. The scenario considered most thoroughly by EPRI assumed that emissions would be capped from now until 2020 and then be required to decline at 3 percent per year starting in 2020, which would produce a 50 percent reduction in emissions by 2050. It also considered two possible technology scenarios: a full portfolio in which all of the technologies considered earlier meet their assumed targets, and a limited portfolio in which carbon capture plus sequestration does not occur and nuclear capacity remains what it is today (Table 11.1). These are "arbitrary

TABLE 11.1 Comparison of Two Possible Technology Scenarios

	Full Portfolio	Limited Portfolio
Supply Side		
Carbon capture and storage	Available	Unavailable
New nuclear	Production can expand	Existing production levels ~100 GW
Renewables	Costs decline	Costs decline more slowly
New coal and gas	Improvements	Improvements
Demand Side		
Plug-in hybrid electric vehicles	Available	Unavailable
End-use efficiency	Accelerated improvements	Improvements

NOTE: The full technology portfolio assumes that all technologies meet their development objectives, while a limited portfolio assumes slower progress. SOURCE: Energy Technology Assessment Center of the Electric Power Research Institute.

assumptions," said Specker, designed to "understand the role of nuclear and coal with and without those resources in the future."

In the full technology portfolio, coal without carbon capture and sequestration phases out by 2040 and is replaced by coal with carbon capture and sequestration (Figure 11.4). Natural gas is used more to meet peak electricity demands than as a baseload source of energy. Consumption is reduced somewhat due to higher prices (as shown by the cross-hatched area at the top of the graph). By 2040, according to this plan, the electricity sector is basically decarbonized, according to Specker. "By 2040 we will have caught up with France in the electricity sector," Specker said, since France already gets most of its electricity from nuclear power and renewable energy sources. "That's always something to keep in mind as we talk about the daunting challenge of decarbonizing the electricity sector—at least one industrialized country has done it."

The situation is very different with the limited technology portfolio (Figure 11.5). According to EPRI's model, to meet the same constraints on carbon dioxide emissions, coal has to be largely phased out by 2040. Reliance on natural gas is much increased. Hydroelectric power, wind power, and other renewables play a much larger role. And the consumption of electricity must be significantly decreased. "You basically are forced to reduce electricity demand because you cannot generate electricity in a low-carbon way." One consequence of such a scenario is that electricity is likely to be much more expensive to dampen demand. Electricity prices could go up an estimated 260 percent to drive down the use of electricity, compared to a 50 percent increase for the full technology portfolio (Figure 11.6).

The cost to the U.S. economy of adopting carbon constraints depends on which technologies are developed (Figure 11.7). With the limited technology portfolio (the left-hand bar on Figure 11.7), the cost of the policy, discounted

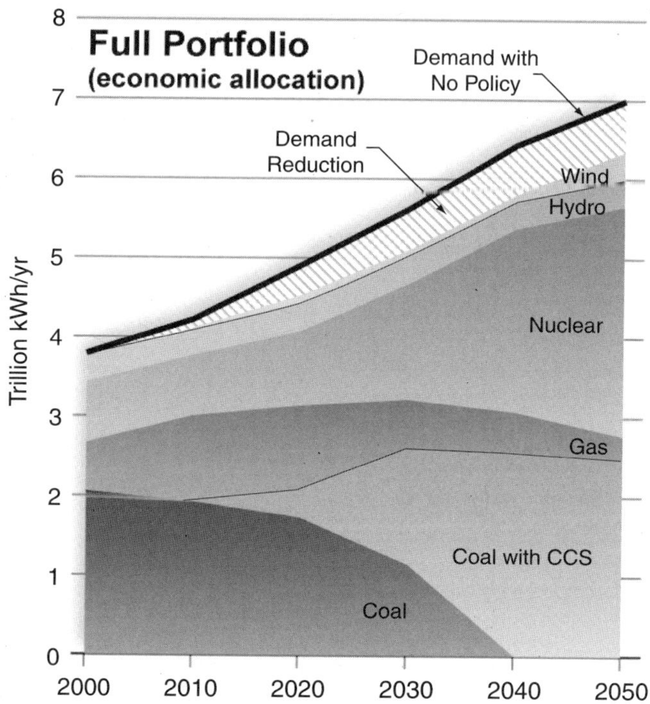

FIGURE 11.4 The full technology portfolio results in the decarbonization of most of the electricity sector by 2040. SOURCE: Energy Technology Assessment Center of the Electric Power Research Institute.

through 2050, is about a trillion and a half dollars, according to the EPRI model's estimates. With the full technology portfolio (the right-hand bar), the cost is about a half trillion dollars. "If we have a carbon dioxide policy in the next few years, which we very likely will, how we then implement that policy with technology is the trillion-dollar question," Specker said. "Technology is critical to managing the cost of a carbon dioxide policy."

For each of the major areas considered in its analysis, EPRI laid out the key technologies that need to be developed to reduce carbon dioxide emissions. These technologies fell into four categories (Figure 11.8). EPRI has created development and deployment roadmaps for each of these technologies showing what is going on now and what will need to be done at various points in the future, "Everyone is very focused about getting things done," said Specker. "We have to get on with [meeting] these challenges." Funding will have to come from the private as well as the public sector. "We're trying to keep the ball in play and keep it moving forward. That's the pragmatic approach."

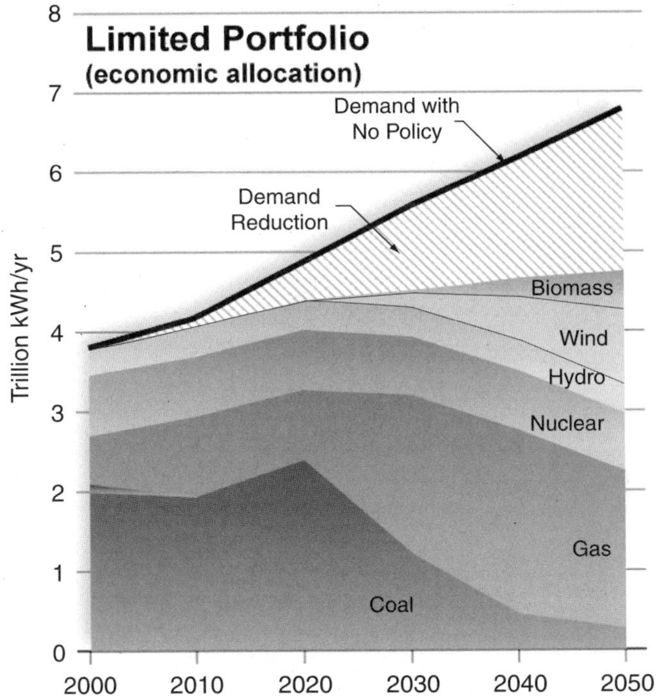

FIGURE 11.5 The limited technology portfolio would require a substantial decrease in electricity use below the business-as-usual scenario. SOURCE: Energy Technology Assessment Center of the Electric Power Research Institute.

REDUCING U.S. GREENHOUSE GAS EMISSIONS

McKinsey & Company, a business consultancy firm that advises corporations and governments, recently conducted a comprehensive analysis of options to reduce greenhouse gas emissions (McKinsey & Company, 2007). The analysis considered both proven, commercialized technologies and four emerging technologies: carbon capture and sequestration, cellulosic biofuels, plug-in hybrids, and LED lighting. It did not examine more speculative technologies in detail. "It's not because we don't believe those will happen," said Ken Ostrowski. "In fact, we're quite encouraged, and we know that as the United States and other economies begin to focus on this task more seriously, there will undoubtedly be important breakthroughs. But we focused our analysis only on those that were proven or the four that I mentioned that were emerging."

The project covered seven sectors of the economy: buildings, power, transportation, industry, waste, agriculture, and forestry. Researchers conducted interviews with more than 100 leading authorities and companies. They also

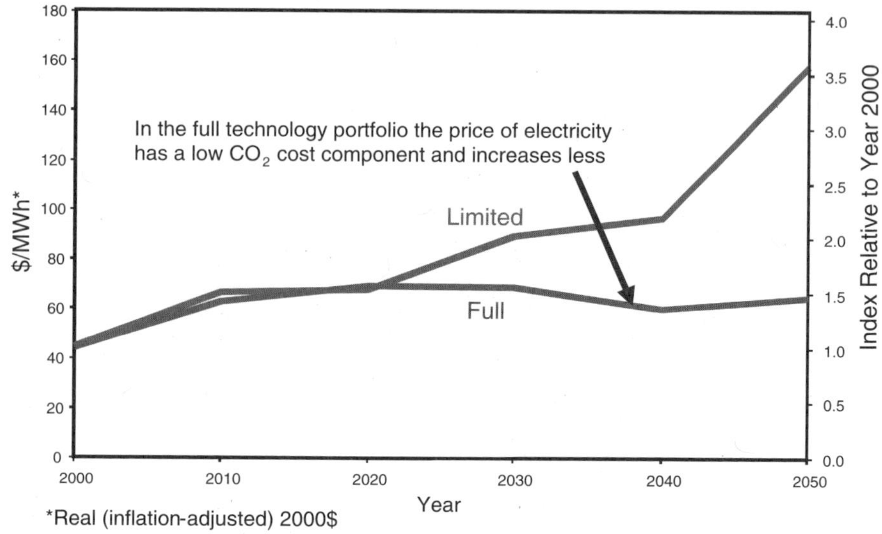

FIGURE 11.6 With the full technology portfolio, the wholesale price of electricity would be much less than with the limited technology portfolio. SOURCE: Energy Technology Assessment Center of the Electric Power Research Institute.

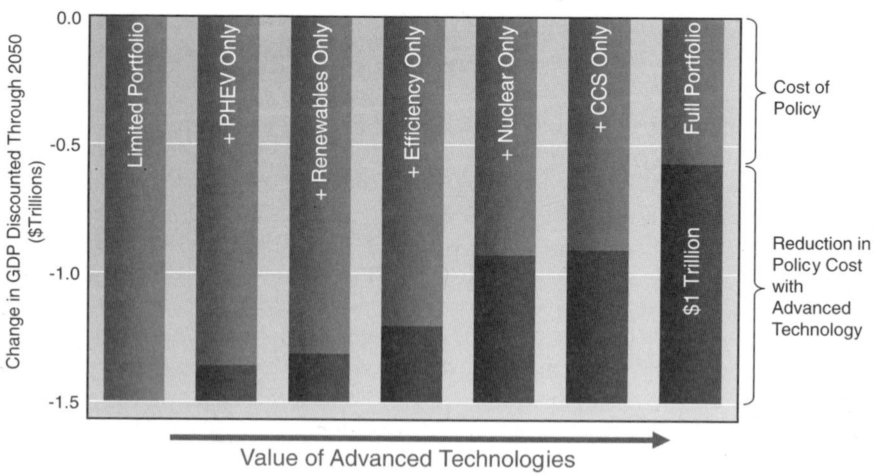

FIGURE 11.7 The change in gross domestic product through 2050 owing to adoption of carbon dioxide reduction policies becomes substantially smaller as more new energy technologies become available. SOURCE: Energy Technology Assessment Center of the Electric Power Research Institute.

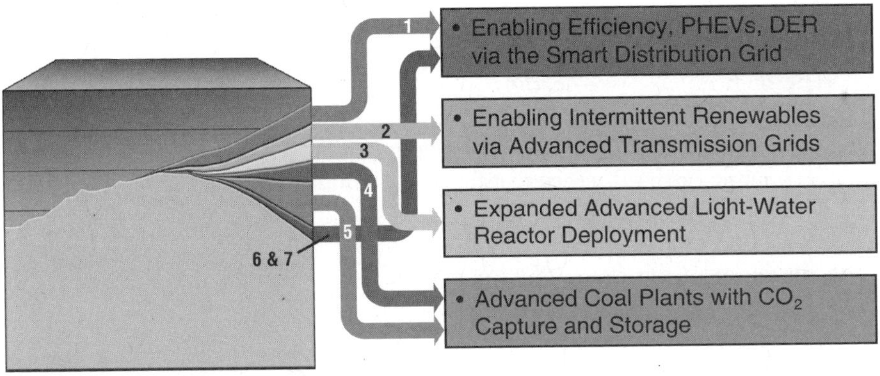

FIGURE 11.8 Reducing carbon dioxide emissions will require technological advances in the four key areas shown on the right. The seven categories of contributing technologies (left) are those shown in Figure 11.3. SOURCE: Energy Technology Assessment Center of the Electric Power Research Institute.

took advantage of the internal expertise available at the company. An academic panel provided support and guidance, and the overall project was sponsored by seven corporate and environmental organizations, although the report remains an independent report put together by McKinsey & Company. "Essentially, we talked to anybody who had expertise and was open to talk with us," said Ostrowski. "We tried to make this a very extensive, comprehensive assessment of the state of knowledge."

Using data from the EIA and other organizations, the McKinsey analysts constructed an emissions reference base from the present to the year 2030. In 2005, the United States emitted approximately 7.2 billion metric tons of carbon dioxide. Under a business-as-usual scenario—with an expanding population, a growing economy, and larger homes and businesses containing more appliances—the expected growth to 2030 was 2.5 billion metric tons, to a total of 9.7 billion metric tons in 2030, a 35 percent increase in emissions. This projection is unlikely to be completely accurate, Ostrowski noted. But it provided a defensible baseline against which to measure emissions reductions.

Based on that projection, the McKinsey project considered three scenarios. In the low-range case, carbon dioxide emissions are 1.3 billion metric tons less in 2030 than in the baseline case. This figure represents a relatively "uncoordinated response to the challenge that the nation faces," Ostrowski said. "Some might say that's the path we're on today, but this essentially says there are incremental improvements over the course that we would have been on otherwise."

The mid-range case, which would result in a 3-billion-metric-ton reduction in emissions, represents a more concerted and coordinated response. This would be "a fairly aggressive response," according to Ostrowski, "but still we would stop shy . . . of saying we took every single option to its maximum economic potential."

The high-range case—leading to a reduction of 4.5 billion metric tons—represents a fully committed response. As Ostrowski described it, this case would imply that "we are absolutely serious about carbon, and we're going to hit every single option that we can to its maximum potential."

The McKinsey report focuses primarily on the mid-range and high-range cases, or a potential abatement of 3 billion to 4.5 billion metric tons. This level is on the order of the reductions called for by various bills that are being discussed in the U.S. Congress. "Only as we get well past our mid-range case and into the aggressive territory do we begin to match the levels that are currently being called for."

The authors of the McKinsey report examined 250 different options for reducing carbon dioxide emissions. They asked how each technology or approach would be developed and commercialized over time, and what level of abatement it could provide. They then aggregated the 250 options into 83 categories and calculated how much abatement each category could provide and the cost of the abatement. The result was a widely reproduced chart (Figure 11.9). The width of each bar on the chart represents the potential abatement attributable to that option in billions of metric tons, with the sum of all the bars about 3 billion metric tons of reduced emissions. "This represents essentially three times the total level of emissions by Germany," Ostrowski said. "Still, even at this level, we would be short of the levels that are being called for today."

The height of each bar represents the cost of that option, with "cost" being the incremental capital, operating, and maintenance costs relative to what would have been spent under the baseline scenario. "If we decide to build an incremental nuclear plant, we would compare the incremental capital, operating, and maintenance costs of that additional nuclear plant relative to what it displaced, which was likely some combination of a supercritical coal plant and a combined-cycle gas plant."

One obvious aspect of the chart is that the set of abatement options is highly fragmented. The widest band represents about 10 percent of the total abatement. "There is no silver bullet," Ostrowski said, "and if one of these options does not deliver the full potential, it does not mean that we cannot achieve near the levels that are projected here."

The second obvious aspect of the chart is that some of the bars extend below the line. These represent options that have a positive net impact on the economy. The incremental capital costs are more than offset by the operating and maintenance savings that are realized over the lifetime of that action. For example, a compact fluorescent bulb has a higher initial cost, but the lifetime and energy efficiency of that bulb are so much greater that use of the bulb results in a net savings. Furthermore, when all the options are considered together, the total savings are approximately equal to the total costs.

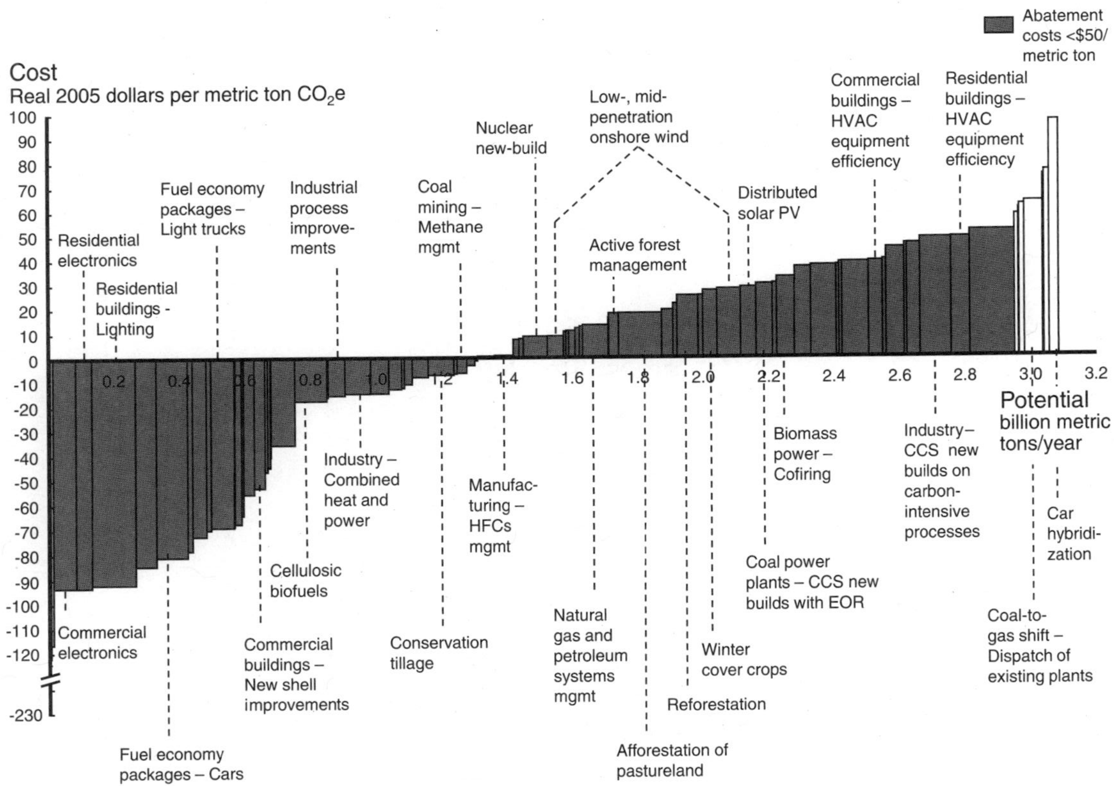

FIGURE 11.9 Eighty-three options for reducing carbon dioxide emissions could result in almost 3 billion metric tons of emissions reductions with net economic benefits (bars below the line) about equal to net costs (bars above the line). SOURCE: McKinsey & Company (2007).

"We would achieve 3 billion metric tons of abatement without incremental costs," Ostrowski said.

Jon Creyts pointed to some of the detailed aspects of the opportunity profile. For example, energy efficiency, which is shown primarily on the left side of Figure 11.9, accounts for roughly 40 percent of the total abatement potential. "Once you change out a light bulb, once you change to a different automobile, once you increase the insulation thickness or put on a reflective roof coating, you have essentially created a durable form of energy efficiency that you can count on," Creyts said.

There are a variety of reasons why the options with positive economic benefit are not necessarily easy to implement. For example, a landlord and a tenant may have competing interests, as may a builder and an owner. Automobile ownership is another issue. The average person owns an automobile for between 4 and 5 years and so does not benefit from the full 14- to 15-year lifespan of a typical automobile. A lack of information also may disrupt or prevent capture of some of the benefits. "Often, our work has been taken out of context, and people use it as a way to push forward the notion that this is cheap and easy to do," Creyts said. "We have said clearly—and we maintain quite clearly—that energy efficiency is very difficult to achieve." However, Creyts added, compared with the challenge of liquefying carbon dioxide gas coming out of the back end of a power plant, pumping it underground, and keeping it there for thousands of years, efficiency improvements deserve special attention.

Ostrowski and Creyts also noted that in many cases policies have to change to enable implementation of emissions-reducing options, and the McKinsey study did not factor in the costs of those policies. "We did not want to prescribe what the policy solution should be," said Ostrowski. "There are many ways to address this issue, and we'll leave that up to the policymakers."

The McKinsey study looked at several categories of technologies with substantial abatement potential (Figure 11.10). In each case, it evaluated the potential under the low-range, mid-range, and high-range cases. For example, with carbon capture and sequestration, the projections started at zero in 2005, with succeedingly higher adoption in each case. In addition, within these categories substantial potential exists for emissions reductions with net benefits to the economy (Figures 11.11 through 11.15).

Ostrowski and Creyts noted that where an option falls on the curve—thus representing its net cost—often depends on the sequencing of events. For example, the rate at which the electric grid or transportation are decarbonized influences a variety of energy efficiency options, such as the use of plug-in hybrids. Also, efficiency improvements could postpone the need to build additional generating capacity until more efficient power plants are developed. If plant construction was delayed, as much as $300 billion of additional investment in generating capacity could be avoided. "If we aren't able to capture that energy efficiency early, we may very well wind up building that additional $300 billion and then idling that capacity in the longer run," said Creyts.

	2005	Low-range case	Mid-range case	High-range case
Coal with CCS (gigawatts)	• 0	22	55	83
Nuclear (gigawatts)	• 100	113	129	153
Renewables (gigawatts)	• Wind – 10 • Solar – <1	70 38	116 80	164 228
Cellulosic biofuels (billion gallons)	• 0	5	14	51
Light-duty vehicle performance (mpg)	• 25 mpg	30	40	44
Efficient new residential lighting	• 8%	15%	70%	75%
Abatement potential below $50/metric ton, in billions of metric tons		1.3	3.0	4.5

FIGURE 11.10 Six categories of advanced technologies could produce low-, medium-, and high-range emissions reductions. SOURCE: McKinsey & Company (2007).

For the mid-range case, the McKinsey study estimates that the up-front capital costs to the economy would be about $1.4 trillion. This amount is only about 1.8 percent of the total real capital investment in the economy over this period, Creyts noted. But it is concentrated in certain sectors of the economy. For example, $560 billion of that investment is within the power sector, which represents "a massive recapitalization of the power sector." Similarly, transportation will have to undergo a significant recapitalization.

Investments in emissions-reducing technologies would have substantial impacts on the energy-producing sector, Creyts observed. Conventional coal-powered energy production would decline substantially, with an increase in carbon capture and sequestration. Energy from renewable sources would increase. Counterintuitively, the use of natural gas declines quite significantly from its current role, which creates a catch 22 for the electricity industry. "If we are unable to capture energy efficiency in the near term, we would wind up building out that gas asset base and wind up essentially idling it in the long-term because gas would compete fundamentally at the margin with renewable power," said Creyts. "That could lead to a large amount of stranded assets here in the United States."

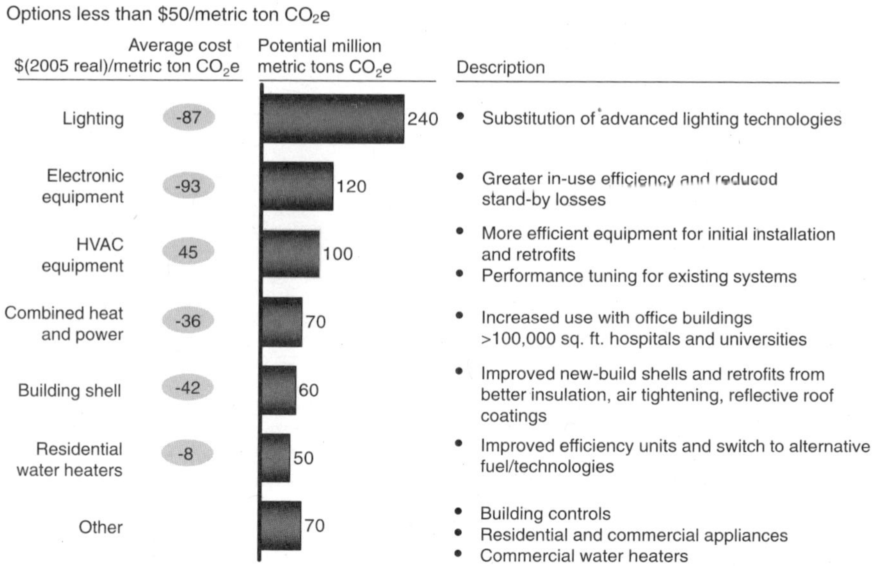

FIGURE 11.11 Improvements in buildings and appliances offer many options with net benefits to the economy. SOURCE: McKinsey & Company (2007).

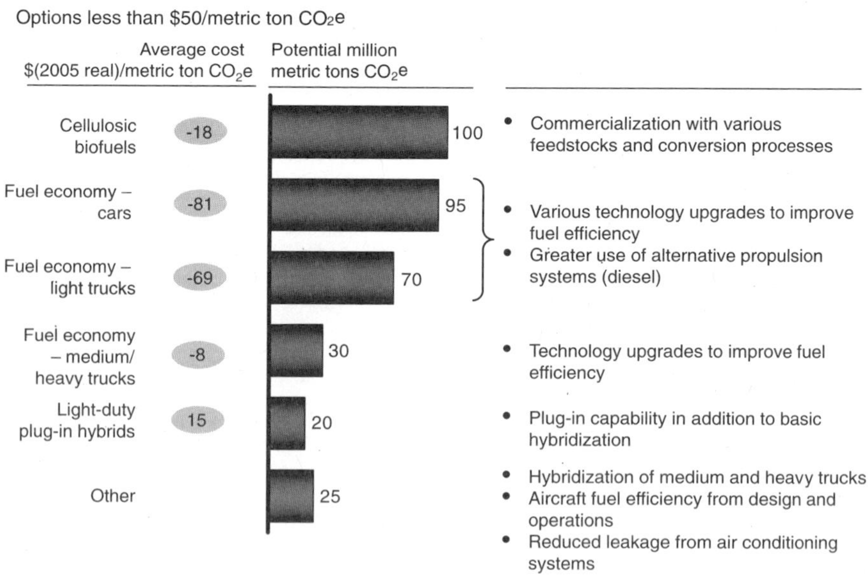

FIGURE 11.12 Vehicle fuel economy and lower-carbon fuels will be essential to reduce transportation emissions. SOURCE: McKinsey & Company (2007).

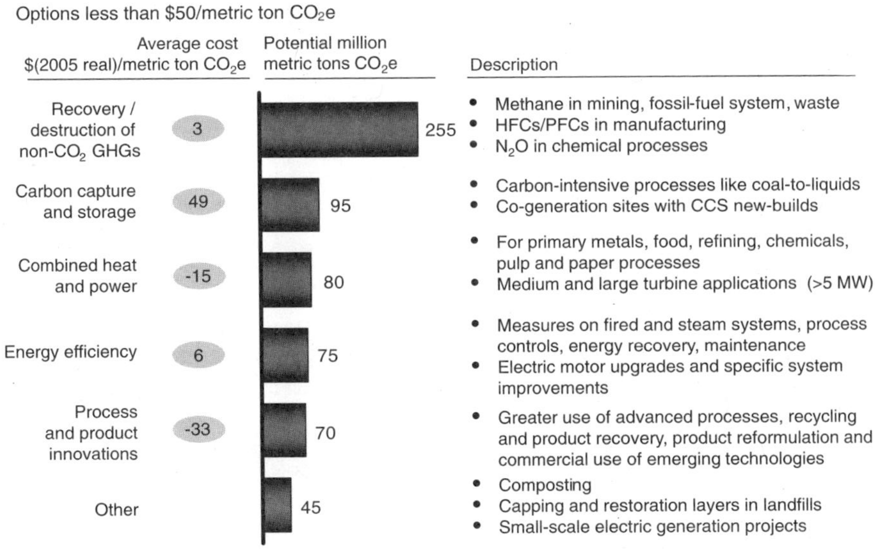

FIGURE 11.13 Options in industry and the waste sector are highly fragmented. SOURCE: McKinsey & Company (2007).

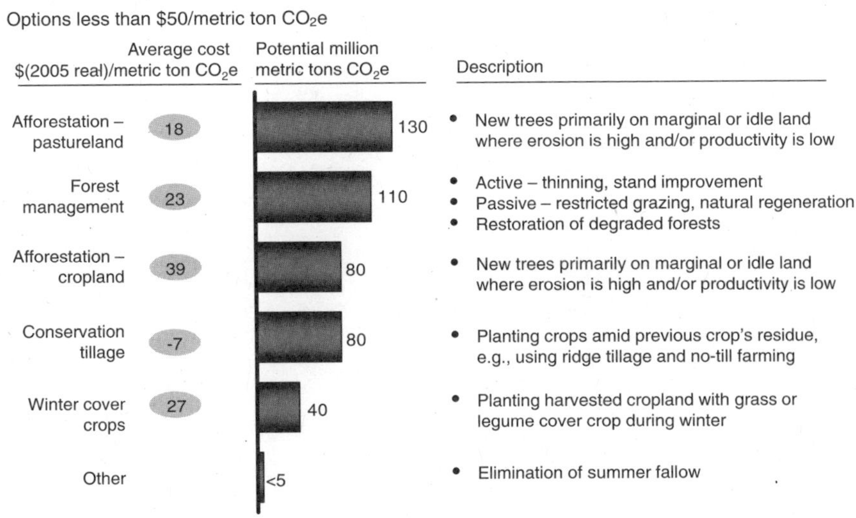

FIGURE 11.14 Terrestrial carbon sinks offer substantial abatement potential at moderate cost. SOURCE: McKinsey & Company (2007).

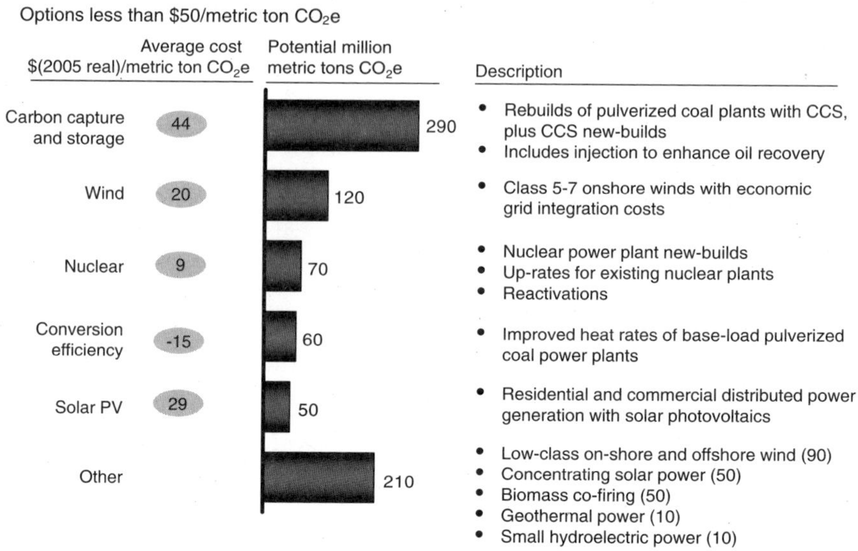

FIGURE 11.15 Electric power generation offers large but higher-cost abatement potential. SOURCE: McKinsey & Company (2007).

12

The Public Sector Response

In December 2004 the National Commission on Energy Policy released its report *Ending the Energy Stalemate* (NCEP, 2004). The Commission, which was launched in 2002 and funded by the William and Flora Hewlett Foundation and other philanthropies, was established as a bipartisan group with members from business, federal and state governments, academia, labor, and non-governmental organizations.

"We started with what clearly was a stalemate in U.S. energy policy," said John Holdren, who co-chaired the Commission with former EPA administrator William Reilly and Exelon CEO John Rowe. A gap between rising oil use and declining domestic production had been widening since 1985, with little policy action to address the gap on either the supply or the demand side. Corporate average fuel economy (CAFE) standards had been unchanged since 1985 for passenger cars and had been constant from 1987 to 2005 for light-duty trucks, including pickups, vans, and sport utility vehicles. The whole-fleet average was 24 miles per gallon in 2003, the same as it had been in 1981. Thirteen years after the United States had ratified the UN Framework Convention on Climate Change, no requirement or incentive was in place to reduce carbon dioxide emissions from the energy sector. No new nuclear reactor had been ordered in the United States since 1978, and the siting of new liquid natural gas terminals and even wind farms had been stymied by NIMBY ("not in my back yard") sentiments, which Holdren said were rapidly transitioning to BANANA sentiments—"build absolutely nothing anywhere near anyone." And federal spending on energy-technology research, development, and demonstration projects in 2004 was the same as in 1987, even though the gross domestic product had almost doubled.

The Commission had two overarching objectives. The first was to develop recommendations that can ensure ample, clean, reliable, and affordable energy for the United States in the 21st century while responding to growing concerns about the nation's energy security and the risks of global climate change. The second was for its recommendations to command the bipartisan support necessary to break the long-running energy policy stalemate in the Congress and be enacted. In seeking to meet these objectives, the Commission adhered to several guiding principles. It preferred market-based solutions and gradual adjustments rather than dramatic interventions. It sought to take into account the law of unintended consequences. It aimed for revenue neutrality, economic efficiency, cost-effectiveness, low consumer impacts, appropriate incentives for future action, flexibility for future adjustments, equity, political viability, and ease of implementation, monitoring, and measurement.

The Commission's 2004 report made a variety of important recommendations. With regard to oil and gas supplies, it encouraged nations with underdeveloped oil reserves to allow foreign investment in their energy sectors. It supported research and development on technologies to mitigate the environmental impacts of developing unconventional oil resources. It urged the United States to fill the Strategic Petroleum Reserve and encouraged other nations to establish publicly owned reserves. It recommended creating incentives for construction of an Alaskan natural gas pipeline and removing hurdles for the siting and construction of liquid natural gas facilities. It also recommended increased resources for public land planning and permitting.

With regard to dampening the growth of demand for liquid fuels, the Commission recommended significantly strengthening federal fuel economy standards for cars and light trucks while also reforming the CAFE program. It urged that manufacturer and consumer incentives be put in place to promote domestic production and increased use of advanced diesel and hybrid-electric cars. It also proposed pursuing efficiencies in the heavy-duty truck fleet and the existing passenger vehicle fleet.

To address climate change risks, the Commission proposed initiating in 2010 a mandatory, economy-wide, tradable-permits system to limit greenhouse gas emissions. The permit system would be designed to reduce the carbon emissions intensity of the United States by 2.4 percent per year and would have a "safety valve" (which is discussed later in this chapter) to prevent excessive economic dislocations. The Commission also recommended linking subsequent U.S. action with comparable efforts by other developed and developing nations via a program review in 2015 and every 5 years thereafter. "This was the provision that was necessary to assure people on the Commission concerned about the competitiveness effect of the United States' embracing a price on carbon emissions and having our major competitors not do so," said Holdren.

A set of recommendations directed to protecting critical energy infrastructure called for addressing the vulnerability of the electricity grid to attack and improving security on cyberattacks against the systems that manage the power grid. In addition, the Commission asked for an examination of whether surveillance technologies developed for defense and intelligence purposes could be applied to widely deployed energy systems.

Research and development on energy technologies received special emphasis in the Commission's recommendations. Its report called for doubling the annual real federal expenditures for energy research, development, and demonstration in the next 5 years, to a level of $3.3 billion in 2004 dollars in 2010. Within this effort, the Commission recommended that funding for international cooperation on energy research, development, and demonstration should be tripled, to $750 million per year, and that the increased spending also should be complemented with a tripling of federal expenditures for accelerated deployment of the most promising technologies that successfully pass the demonstration phase. These expenditures would amount to $2 billion per year by 2010. Finally, the Commission concluded that the tax code should be revised to increase private-sector incentives to invest in energy research, development, demonstration, and early deployment.

Cleaner coal technology should be one focus of the technology-innovation effort, Holdren said. "I use that term deliberately rather than 'clean coal technology' because many people point out there is no such thing as clean coal technology, but we can certainly make it cleaner." Specifically, the Commission recommended that federal early-deployment incentives of $400 million over the next decade should be directed to faster commercialization of integrated gasification combined-cycle coal plants, which sharply reduce emissions of air pollutants, produce liquid and gaseous fuels as well as electricity, and can be more easily retrofitted to capture carbon dioxide. Additionally, it concluded that the development and commercial-scale demonstration of carbon dioxide capture and sequestration technologies also should receive $300 million in federal support over the next decade.

The Commission called for nuclear energy technologies to receive $2 billion over the course of a decade from the federal government for "first mover" advanced nuclear power plants to demonstrate improved safety and economics. The United States also should move expeditiously to establish a project for centralized, interim storage of spent fuel at no fewer than two locations. "We would no longer have all the eggs in the Yucca Mountain basket," Holdren said. In parallel, the United States should work to reduce the risk of nuclear proliferation by reiterating a commitment to continue indefinitely the longstanding moratorium on commercial reprocessing of spent nuclear fuel and the construction of commercial breeder reactors, emphasizing the policy of discouraging the accumulation of separated plutonium in civil fuel cycles elsewhere, and working to prevent the deployment of uranium-enrichment and spent-fuel-reprocessing capacity in additional countries. "Some of these might be seen as controversial,

[but] all of these recommendations were unanimous. There were no dissenting opinions in this diverse and bipartisan group about any of these matters."

In the area of renewable energy technologies, the Commission said, the United States should accelerate the development and deployment of non-petroleum transportation fuels, especially cellulosic ethanol and diesel from biomass and wastes. Research, development, and demonstration should be increased from $25 million to $150 million per year over 5 years, and $750 million in early deployment incentives should be funded from 2008 to 2017, according to the Commission. In addition, research, development, and demonstration on solar photovoltaic and solar thermal energy systems should go from $83 million to $300 million per year. And the renewable energy production tax credit should be extended and expanded to include all energy sources that do not emit carbon.

The Commission recommended increasing manufacturer and consumer incentives for more efficient vehicles from $80 million per year in 2004 to $300 million per year. It also called for increasing federal research, development, and demonstration funding on efficiency improvements in buildings and appliances from $60 million to $300 million per year. And funding on improved efficiency in industrial processes should go from $93 million per year to $200 million per year.

Both the Commission and the Energy Information Administration conducted an analysis of the economic impact of its recommendations. According to the EIA, the impact of the carbon emissions permit system would not exceed 0.15 percent of the gross domestic product in 2025. "At the forecasted rate of growth—2.8 percent—Americans would have to wait until about January 18, 2025, to be as rich as they otherwise would have been on January 1 of that year," Holdren pointed out. Furthermore, the full set of policies recommended by the Commission would reduce the gross domestic product by no more than 0.4 percent in 2025 while reducing greenhouse gas emissions by 11 percent from the reference case. However, although coal use in 2025 would be 10 percent below the reference case, it would still be 22 percent above the level of 2003.

ACTIONS TAKEN BY THE BUSH ADMINISTRATION

Reuben Jeffrey laid out the actions taken by the Bush Administration in general, and the State Department in particular, to address issues of energy security and the environment. First, the administration has emphasized diversification away from hydrocarbons over the medium and long term. For example, a week before the Academies' energy summit, the U.S. government hosted the Washington International Renewable Energy Conference to highlight the importance of renewable and alternative energy technologies.

Since 2001, ethanol production has quadrupled from 1.6 billion gallons to an estimated 6.4 billion gallons in 2007, Jeffrey said. Biodiesel production is up 80 percent from 2006. Wind energy production has increased by more than 300 percent from 3 years ago, and solar capacity has doubled in the past several years. Admittedly, all of these increases are from small bases, and each source needs to develop more quickly to become a significant element of energy supplies, Jeffrey acknowledged, but the growth has been significant.

The State Department is helping governments, private companies, and researchers collaborate on promising technologies, Jeffrey said. For example, a partnership between the United States and Brazil is intensifying collaborative research to speed the commercialization of the next generation of biofuels and catalyze sustainable production of biofuels in countries in the Western Hemisphere. The State Department also is developing compatible biofuels standards in the Western Hemisphere and in Europe—a necessary step for biofuels to become a global economic commodity.

Since 2001 the administration has dedicated and the American taxpayer has invested $37 billion in science and technology research related to climate change, including $18 billion for the development and promotion of clean energy technology, Jeffrey said. For example, President Bush has made a $2 billion commitment to a clean technology fund administered by the World Bank and supported by Japan, the United Kingdom, and other partners. The fund aims to bring the best available clean energy technologies to emerging markets.

International partnerships initiated or led by the United States are working to bring clean, safe civilian nuclear power to developing countries, Jeffrey observed. These partnerships increase the security of energy supplies, allow countries to become less dependent on foreign oil and gas, and limit the spread of potentially dangerous weapons technologies.

In 2006, at the St. Petersburg Summit, the G-8 nations agreed to various principles that address energy security, investment in the energy sector, sustainable development, and climate change. These principles include support for open, transparent, efficient, competitive energy markets; diversification of energy sources and routes; and environmentally sound development and use of energy. These principles are very much in keeping with U.S. domestic policy goals, Jeffrey observed.

In Europe and Asia, the United States is working with regional partners and private companies to encourage increased energy production and greater diversity in transit routes to bring these products to market. The United States is partnering with Asian countries through the Asia-Pacific Economic Cooperation forum to improve energy efficiency, reforestation, and cooperation in green technology. A formal strategic economic dialogue with China, and broader energy dialogues with China and other emerging economies, seek to encourage the adoption of market-based energy policies, the rapid adoption of clean energy technologies, and a responsible approach to the development of oil resources.

The United States also has launched multilateral technology initiatives such as the International Partnership for the Hydrogen Economy, which explores the advancement of hydrogen as a fuel, and the Carbon Sequestration Leadership Forum, which works to improve technology to capture carbon dioxide and store it safely underground. The Methane to Market Partnership seeks to capture and use the greenhouse gas methane as a fuel source instead of releasing it into the atmosphere.

To directly address global greenhouse gas emissions, the United States is committed to developing an environmentally effective and economically sustainable framework under the UN Framework Convention on Climate Change, Jeffrey stated. At the UN Climate Conference in Bali in December 2007, the United States helped forge consensus on a roadmap for these negotiations, which are scheduled to conclude by December 2009. To advance these negotiations, President Bush launched the Major Economies Process to bring together the top energy-consuming countries from the developed and developing world, which together represent some 80 percent of the world's energy use, economic growth, and greenhouse gas emissions. Through the Major Economies Process, the United States hopes to build consensus among the key players in a number of areas, including a shared long-term global emissions reduction goal, national mid-term plans and goals, and cooperative technology strategies in key sectors.

Solving U.S. energy problems will require many years, Jeffrey observed. Accordingly, the United States must take action today and do more to improve energy security and address the challenges of global warming. At the same time, the nation must confront its continued reliance on oil, natural gas, and coal, which creates political, economic, and environmental challenges. Cooperation among governments and the continued dedication of many individuals and groups will be essential for success, Jeffrey said.

At the Department of Energy, the administration has boosted investments in research and development at all stages of the innovation cycle to help the United States break its "over-dependence on fossil fuels," according to Samuel Bodman. At a very broad level, President Bush has proposed a linked set of increases for federally funded research in the physical sciences under the American Competitiveness Initiative. "This is serious money for serious science in areas like supercomputing, nanotechnology, advanced nuclear reactor technologies, and fusion energy," Bodman said. "The results may not be seen for 5 or 10 years, or even decades, but the critical investments must be made now." The $4.7 billion request for the Department of Energy's Office of Science in fiscal year 2009, an increase of almost 20 percent over the enacted fiscal year 2008 appropriation, reflects the administration's commitment to sustaining vital investment in the physical sciences, Bodman said. "Getting Congress to actually appropriate at the levels it has authorized and that the President has committed to has been, let us say, a challenge," Bodman observed. "But we are hopeful, because the commitment to science stems not from a sense that there are immediate political gains to be had from such funding, but from a deep recognition that our energy future rests on sustained leadership in basic research."

The administration also has laid out an aggressive strategy to expand the availability of renewable energy and alternative fuels, Bodman said. The president's Advanced Energy Initiative is identifying the technologies that could have the greatest impact on the marketplace in the relatively near future and then pursuing those technologies with increased resources and aggressive timelines. Examples include cellulosic biofuels, advanced hybrid vehicle technologies, hydrogen fuel cells, solar photovoltaics, and high-efficiency wind power. "These are things that are already in the pipeline and, as a matter of sound public policy, need to be pushed more quickly to market," Bodman said. Bringing these technologies to market will require collaborations among government, industry, and academia, and the federal government is using a range of collaborative models—including cost-sharing partnerships and loan guarantee programs—to share with the private sector some of the risk of developing commercially viable, innovative technologies. For example, Bodman said, six large-scale biorefinery projects together will receive up to $385 million—and a total of more than $1.2 billion—through public-private partnerships over the next 4 years. When fully operational, these six biorefineries are expected to produce more than 130 million gallons of cellulosic ethanol per year.

In addition, the Office of Science is investing more than $400 million over 5 years in three cutting-edge Bioenergy Research Centers. These centers are attracting world-class scientists and engineers from academia, industry, and national laboratories to bring the latest tools of the biotechnology revolution to bear on clean energy production. A major focus of the centers will be understanding how to reengineer biological processes to develop new, more efficient methods for converting the cellulose in plant material into ethanol or other biofuels that serve as a substitute for gasoline. After 6 months of work, promising scientific results are already emerging from this investment, Bodman said.

Ray Orbach elaborated on the money being distributed by the Office of Science. Over the next 5 years, the office is supporting 20 to 30 Energy Frontier Research Centers, which can be located in universities, national laboratories, or the private sector. Each will be funded at a level of $2 million to $5 million for an initial 5-year period. "We want to bring the best talent in our country . . . to, literally, save our Earth," Orbach said. "There is a continuum of investment that this country has to make. We can't guarantee that the investment will work, but we can guarantee that if we don't make the investment, we're stuck with last century's technologies and we won't get there."

Bodman noted that he had issued a policy statement that laid out guiding principles, responsibilities, and a review process to ensure that new technologies are deployed and that continuity and uniformity of technology transfer activities are maintained throughout the Department of Energy. Also, three venture capital firms were recently selected to participate in the department's Entrepreneur in Residence pilot program, which aims to accelerate deployment and commercialization of advanced clean energy technologies from three national laboratories into the global marketplace. "By empowering researchers and entrepreneurs, we are furthering President Bush's initiatives aimed at developing and deploying cutting-edge technologies to address the challenges that face our nation," Bodman stated.

Even as the administration has been emphasizing renewables and alternative fuels, it has recognized that the U.S. economy will remain heavily dependent on fossil fuels. Efficiency and the reduction of carbon emissions—including the demonstration of carbon sequestration capacity—therefore remain priorities at the Energy Department, Bodman said. The department has funded four cost-shared carbon sequestration projects in the United States and plans to fund three more. These projects will conduct large-volume tests for the storage of 1 million or more tons of carbon dioxide in deep saline reservoirs. Collectively, these formations have the potential to store more than 100 years' worth of carbon dioxide emissions from all major sources of pollution in North America.

Access to safe and emissions-free nuclear power also must be expanded in the United States, Bodman observed, while responsibly managing waste and dramatically reducing proliferation risks. Nuclear power is currently the only mature technology that can supply large amounts of emissions-free baseload power to help meet the expected growth in consumption. The federal government has not licensed construction of a new nuclear plant in the United States in nearly 30 years, Bodman noted. "That must change. We are working to see that it does by, among other things, implementing federal risk insurance, or so-called 'stand-by support,' and loan guarantees to try to remove some of the roadblocks associated with getting the next generation of nuclear plants on line." With the rest of the world on the verge of a major nuclear expansion, President Bush introduced in 2006 the Global Nuclear Energy Partnership (GNEP) to facilitate the worldwide expansion of nuclear energy for peaceful purposes in a safe and secure manner. "This historic partnership continues to expand," Bodman said. "Last month, the Department of Energy welcomed the United Kingdom as the GNEP's twenty-first partner."

Robert Marlay described the U.S. Climate Change Technology Program at the Department of Energy, which coordinates research and development across 10 federal agencies. The program has four strategic goals related to emissions: reduce emissions from energy end use and infrastructure, reduce emissions from energy supply, capture and sequester carbon dioxide, and reduce emissions from non-carbon-dioxide gases. It also has two cross-cutting, supporting strategic goals: improve capabilities to measure and monitor greenhouse gases, and bolster basic science and strategic research. The total budget request for the program for fiscal year 2009 was $4.4 billion.

The plan is focused on the United States but is set within a global context,

Marlay said. It reflects the extent to which carbon dioxide emissions must fall to stabilize the concentration of carbon dioxide in the atmosphere at particular levels (Figure 12.1).

To achieve a "net-zero emissions future," progress must be made on all four of the program's strategic goals (Table 12.1). The underlying message is that "all four of those goals are musts," according to Marlay. "We can't just focus on efficiency. We can't just focus on supply. We've got to do something in sequestration. And, perhaps surprisingly, for those of us who spend most of our life in the energy world, other greenhouse gases are a very major contributor."

Two particular areas stand out in the effort to achieve net-zero emissions. The first is the need to decarbonize the electricity grid through such measures as nuclear power, low-emissions coal power, and renewable power. The other is to "de-oil" transportation, using such measures as hybrid and electric vehicles, alternative fuel vehicles and bio-based fuels, and alternative forms of transportation. Once the grid is decarbonized, moving transportation demands onto the grid is a way to substantially reduce emissions. According to Marlay, the Climate Change Technology Program is pursuing these needs through "novel concepts, interdisciplinary concepts, concepts that are cross-cutting and go across the different stovepipes of different agencies."

The agencies represented by the Climate Change Technology Program have hundreds of separate activities in place that are focused on the program's broad goals. A portfolio analysis done by the program has sought to prioritize these activities by identifying technologies that are ready to have a major impact on the problems. By bringing on technologies earlier, cumulative emissions can be substantially reduced, said Marlay. "You're starting earlier, and you get on that path much quicker. . . . What we need to do is we need to figure out how to . . . craft a portfolio that brings it on much sooner, and gives us a much larger benefit."

Even in a very demanding budget environment, the Climate Change Technology Program has been receiving increased budget requests and appropriations. Furthermore, these increased investments serve a dual purpose, Marlay said, in that most of the investments made to reduce the potential of climate change also result in greater energy security. For example, if oil intensity is measured by barrels of oil used per trillion dollars of gross domestic product, technology development leading to more efficient use of oil as well as replacement of oil with alternatives can reduce oil intensity substantially below a business-as-usual case. When oil intensity is high, said Marlay, rapid increases in oil prices can cause economic havoc. At the time of the 1973 oil embargo, U.S. oil intensity was high, as was the economic disruption caused by the event. In contrast, during the recession in the late 1980s, oil intensity in Japan was much lower, even though Japan imports 100 percent of its oil. As a result, Japan experienced less economic disruption during that recession than did other countries. If we can reduce oil intensity through technology development, Marlay said,

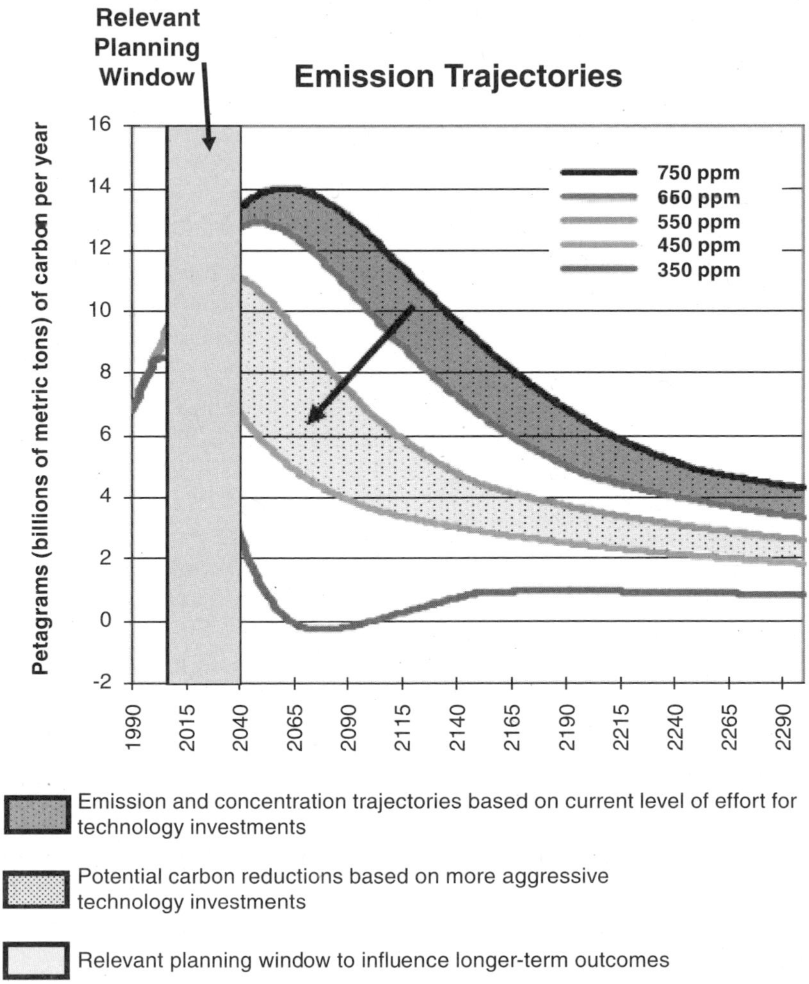

"we're going to be very minimally exposed to price shocks because I believe that oil will largely vanish from the economy."

Advanced technologies could significantly reduce the costs of imposing constraints on carbon dioxide emissions. Very high constraints on greenhouse gas levels—somewhere on the order of 450 or 500 parts per million—have a global cost around $250 trillion (in undiscounted dollars) over the 21st century if technologies evolve at the rate they are today. If technology development could be accelerated, the cost could go down 50 to 70 percent, according to Marlay. This would be a huge benefit, said Marlay, that "is well worth going after." Timing is also critical in reducing the stabilization level of greenhouse

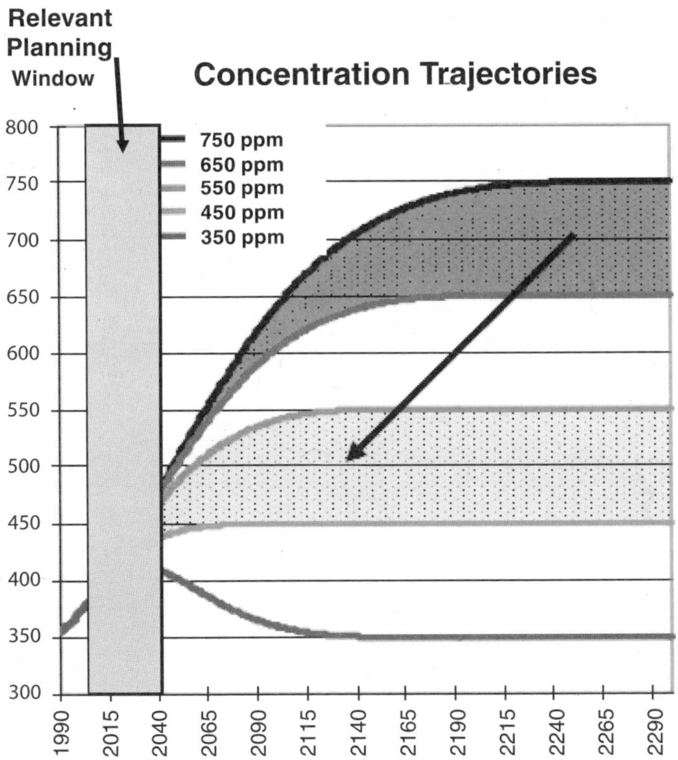

FIGURE 12.1 Investments in technology are needed to reduce emissions (facing page) and lower the level at which carbon dioxide concentration is stabilized (above). SOURCE: Adapted by DOE from Wigley et al. (1996; Figure 1).

gases. For the 450- to 500-ppm scenarios, each of the goals of the Climate Change Technology Program would need to be achieved more quickly than under less severe constraints. "We can't wait forever," Marlay said.

Marlay discussed three potential barriers to achieving the program's goals, which apply to all of the scenarios he and others described at the meeting. The first is how to increase the level of research and development funding made available by national governments. "I'm very pleased, if not proud, of the federal government's leadership on these technologies," he said. "Japan is the only other country that really is devoting the same kind of resources to this particular problem. All the other countries are dabbling." However, Marlay noted, the

TABLE 12.1 A Roadmap for Climate Change Technology Development Provides Near-Term, Mid-Term, and Long-Term Options

	Near-Term	Mid-Term	Long-Term
Goal 1 Energy end-use and infrastructure	Hybrid and plug-in hybrid electric vehicles Engineered urban designs High-performance integrated homes High-efficiency appliances High-efficiency boilers and combustion systems High-temperature superconductivity demonstrations	Fuel cell vehicles and H_2 fuels Low-emission aircraft Solid-state lighting Ultra-efficient heating, ventilation, air conditioning, and refrigeration "Smart" buildings Transformational technologies for energy-intensive industries Energy storage for load leveling	Widespread use of engineered urban designs and regional planning Energy managed communities Integration of industrial heat, power, process, and techniques Superconducting transmission and equipment
Goal 2 Energy supply	IGCC commercialization Stationary H_2 fuel cells Cost-competitive solar PV Demonstrations of cellulosic ethanol Distributed electric generation Advanced fission reactor and fuel cycle technology	FutureGen scale-up H_2 co-production from coal/biomass Low-wind-speed turbines Advanced biorefineries Community-scale solar Gen IV nuclear plants Fusion pilot plant demonstration	Zero-emission fossil energy H_2 and electric economy Widespread renewable energy Bio-inspired energy and fuels Widespread nuclear power Fusion power plants

Goal 3 Capture, storage, and sequestration	CSLF and CSRP Post-combustion capture Oxygen-fuel combustion Enhanced hydrocarbon recovery Geologic reservoir characterization Soils conservation Dilution of direct injected CO_2	Geologic storage proven safe CO_2 transport infrastructure Soils uptake and land use Ocean CO_2 biological impacts addressed	Track record of successful CO_2 storage experience Large-scale sequestration Carbon- and CO_2-based products and materials Safe long-term ocean storage
Goal 4 Other gases	Methane to markets Precision agriculture Advanced refrigeration technologies PM control technologies for vehicles	Advanced landfill gas utilization Soil microbial processes Substitutes for SF_6 Catalysts that reduce N_2O to elemental nitrogen in diesel engines	Integrated waste management system with automated sorting, processing, and recycle Zero-emission agriculture Solid-state refrigeration and air conditioning systems
Goal 5 Measure and monitor	Low-cost sensors and communications	Large-scale, secure data storage system Direct measurement to replace proxies and estimators	Fully operational integrated measure and monitor systems architecture (sensors, indicators, data visualization and storage, models)

SOURCE: CCTP (2006).

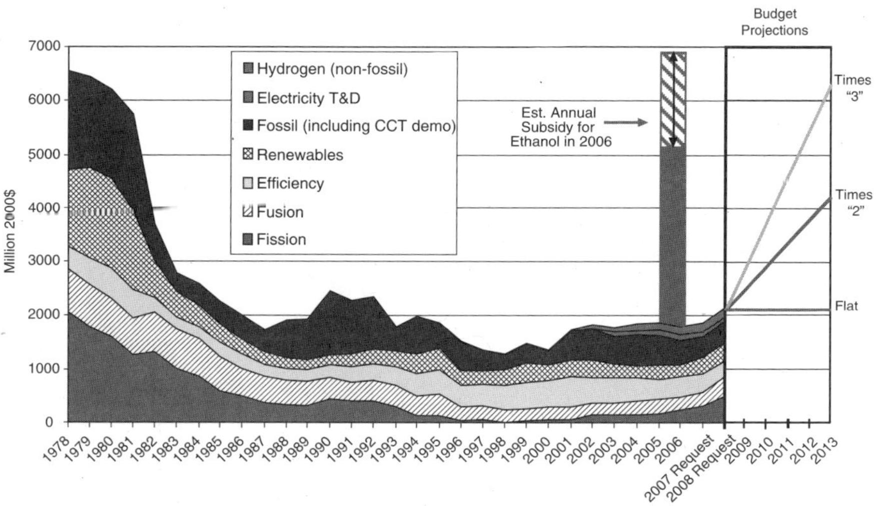

FIGURE 12.2 U.S. Department of Energy energy research, development, and demonstration (RD&D), FY1978-FY2008 administration request. Support for RD&D from the Department of Energy could rise to 1978 levels if funding were to triple. NOTE: The order in the key from bottom to top matches the order of the categories of energy RD&D displayed from bottom to top. SOURCE: Kelly S. Gallagher, Energy Technology Innovation Project, Belfer Center for Science and International Affairs, Kennedy School of Government, Harvard University, available at http://belfercenter.ksg.harvard.edu/publication/18152/doe_budget_authority_for_energy_research_development_and_demonstration_database.html.

U.S. government has the capacity to do much more relative to other countries, including doubling or tripling the Department of Energy's support for energy research, development, and demonstration (Figure 12.2).

The second potential barrier is how to send the proper price signals to the private sector to generate investments in technology. "If you set [a price] too low, . . . you're not going to get the experimentation that you need. If you set it too high, obviously there can be damage to the economy, unless you're very, very clever about recycling the funds back into the economy—ideally back to the payer at some point."

The third barrier is how to advance international collaboration and partnering, and Marlay concluded by discussing the international dimensions of climate change. Global participation will be essential to control climate change. That participation will require realistic goals and commitments, which will require careful negotiations among developed and developing countries. "There's a debate between the North and South and the developing nations and the developed nations," Marlay observed. The developing nations say, "You've feasted for 200 years. Now, we've just come to the table and you want us to go on a diet. How is that fair?" However, there is also a counterargument, Marlay noted, which was offered by the Canadians at the recent Bali conference. "Yes, but we're all on a lifeboat and we're sinking. Don't you want to help to bail?"

"So these are the challenges," Marlay said. How can global investments in research and development be increased? How can private sector innovation be enhanced? How can international cooperation be advanced? "I have been excited about what I have heard over the past 2 days," Marlay said.

ACTIONS TAKEN BY CONGRESS

Senator Jeff Bingaman described the "significant actions" that have been taken in the current Congress and the previous Congress in response to the challenges of energy supply and use. The Congress passed a major energy bill in 2005, when Republicans controlled the Congress, and another in 2007, when Democrats controlled the Congress, both by large margins. In particular, the 2007 Energy Independence and Security Act took six steps that Bingaman considers turning points in energy policy.

1. It mandated the first increase in 32 years in statutory CAFE standards for both cars and trucks.
2. It called for the use of biofuels to grow to 36 billion gallons in 2022, with subsidiary targets within that total for cellulosic ethanol, biodiesel, and other advanced biofuels.
3. It required improved efficiency standards in the use of energy, with a special focus on buildings and lighting.
4. It strengthened the federal commitment to energy research and technology development, including the Department of Energy's program for carbon capture and geological storage.
5. It authorized a strong "green jobs" training program, since a large segment of the current energy workforce will soon be eligible for retirement.
6. It created new protections for consumers against manipulation in oil and gas markets.

The EIA has calculated that the 2007 bill will reduce the nation's dependence on imported oil and will slow the growth of energy-related carbon emissions, Bingaman noted. Compared to the baseline before the law's enactment, oil imports continue to decline after 2010, according to this analysis, and they stay significantly lower than that baseline. By 2030, imports are reduced by more than 2 million barrels per day from the level that would have been expected without the legislation (Figure 12.3). Energy-related emissions of carbon dioxide also are forecast to decrease by 500 million metric tons as a result of the new act—an amount equivalent to the annual carbon dioxide emissions from South Korea, the world's ninth largest source of carbon dioxide emissions (Figure 12.4).

However, "compared to where we need to go, these are still very modest steps," Bingaman said. Carbon dioxide emissions will still grow, and oil imports will still be substantial. "Clearly, there is much more that needs to be done."

Nevertheless, as Holdren pointed out, the 2005 legislation did break the stalemate that the National Commission on Energy Policy had been established to address. "It embraced most of the Commission's recommendations on a number of topics—on oil and gas supply, on energy infrastructure, on the incentives for deployment of renewables, nuclear, and clean-coal technology, and on R&D incentives for industry."

There were three main recommendations from the Commission (NCEP, 2004) that the 2005 legislation failed to address, Holdren observed. It did not strengthen the CAFE standards (although the 2007 energy legislation did take that step). It did not significantly increase federal energy research and development. And it did not establish mandatory economy-wide greenhouse gas restraints.

Regarding the third point, a Sense of the Senate Resolution passed on June 22, 2005 (see *Congressional Record—Senate*, June 22, 2005), was interesting for what it said about climate change, according to Holdren. It read:

> It is the sense of the Senate that Congress should enact a comprehensive and effective national program of mandatory, market-based limits and incentives on emissions of greenhouse gases that slow, stop, and reverse the growth of such emissions at a rate and in a manner that (1) will not significantly harm the United States economy; and (2) will encourage comparable action by other nations that are major trading partners and key contributors to global emissions. (p. S7089)

The resolution did not get the vote of every Democrat, Holdren observed, but it got eight votes from Republicans, which was enough for it to pass.

The National Commission on Energy Policy remained in existence, with somewhat different membership, after releasing its 2004 report (NCEP, 2004) and released another major report in April 2007 (NCEP, 2007). The updated recommendations again clustered in several areas. The report recommended a 4 percent per year improvement in CAFE standards, with the National Highway Traffic Safety Administration authorized to modify the target up or down. It also called for other cost-effective reductions in transportation energy use, with a focus on heavy-truck fuel economy and efficiency standards for light-duty vehicle replacement tires.

To address climate change, the Commission again called on Congress to implement a mandatory market-based program to limit economy-wide U.S. greenhouse gas emissions. This time the targets were more ambitious than in the 2004 report—to return to 2006 levels by 2020 and to get 15 percent below that level by 2030. The 2007 NCEP report also called for a "safety valve" price

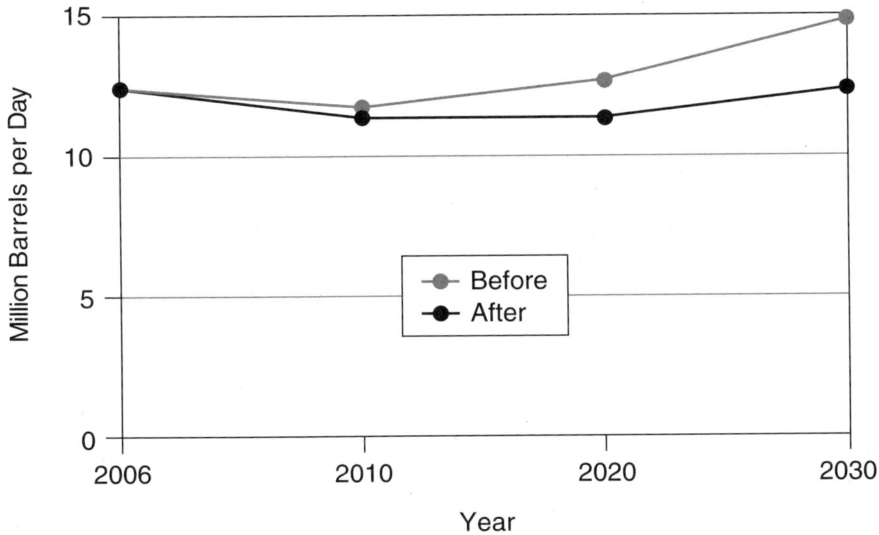

FIGURE 12.3 The 2007 Energy Independence and Security Act is projected to reduce the amount of oil imported into the United States by more than 2 million barrels per day in 2030. SOURCE: EIA (2008).

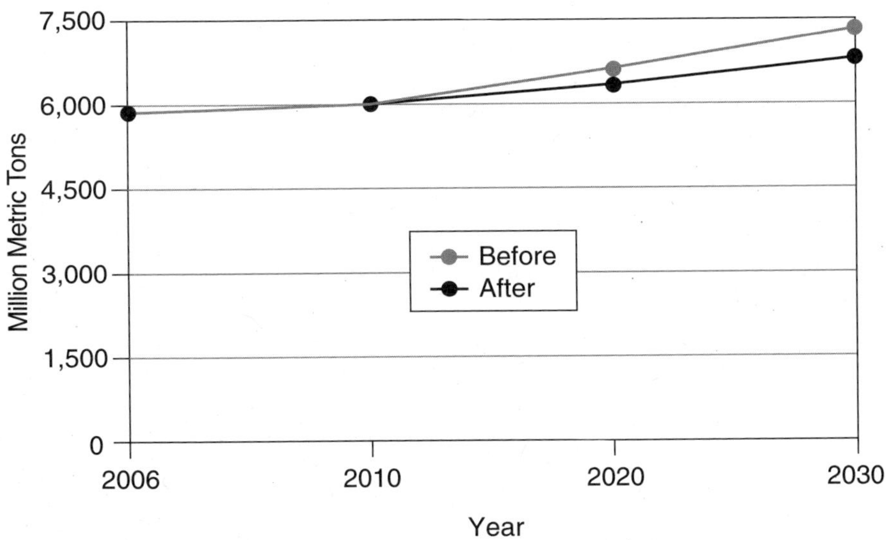

FIGURE 12.4 The 2007 Energy Independence and Security Act is projected to reduce U.S. carbon dioxide emissions by approximately 500 million metric tons by 2030. SOURCE: EIA (2008).

of $10 per ton of carbon dioxide, escalating at 5 percent per year in real terms. Half of the permits would be distributed to affected industries. The rest would be auctioned and used to increase incentives for advanced technologies and to reduce impacts on low-income individuals.

The Commission called for incentives for carbon capture and sequestration using bonus allowances that would be at least equal in value to renewable production tax credits. It also called for stronger incentives for comparable action on the part of key trading partners. The United States should provide technical and financial resources for the transfer of low-carbon technologies, signal its determination to address trade and competitiveness concerns, and link future commitments by the United States to international progress, the Commission said.

On efficiency, the 2007 report called for enhancing and extending the tax incentives that were created or extended under the 2005 act. It also called on the Department of Energy to follow through on issuing efficiency standards for 22 categories of appliances and equipment.

Regarding renewable sources of energy, the Commission recommended extending the eligibility period for federal production tax credits in 5-year rather than 1- or 2-year increments to provide certainty for the industry that those tax credits would continue to exist. Perhaps most controversially, the new report called for a federal renewable portfolio standard that would increase the share of electricity generated by renewable sources to at least 15 percent by 2020. This was a step the initial Commission had not recommended.

The 2007 NCEP report repeated the major recommendations of the earlier report on natural gas and coal. It also recommended conditioning eligibility for public funding or subsidies on the inclusion of carbon capture and sequestration for any new advanced coal projects. "In other words, saying that something is going to be carbon-capture ready would not be good enough," Holdren said. "In order to qualify for public subsidies, the advanced coal projects would actually have to be doing it." In addition, new coal plants built without carbon capture sequestration should not be grandfathered under future greenhouse gas regulations. The Environmental Protection Agency should complete as soon as possible a rigorous, formal, and public process to formulate effective regulatory protocols governing long-term storage of carbon dioxide. And carbon capture and sequestration should be included from the outset in any taxpayer-supported efforts to develop coal-to-liquids technology. "This is fairly hard-hitting stuff, if you think about it, and again it was unanimous," Holdren said.

The 2007 NCEP report called more explicitly for amending the Nuclear Waste Policy Act to align its requirements with human engineering and scientific capabilities, while adequately protecting health and environment. It said that the Department of Energy should site and operate consolidated national or regional interim storage options, take possession of and/or remove fuel from reactor sites that have been or are being decommissioned, and support research and development on alternatives to geological disposal of spent fuel. Plans for interim storage and the federal responsibility for disposal should be sufficient to satisfy the Nuclear Regulatory Commission's waste confidence requirement.

The Commission called for re-evaluating ethanol subsidies and tariffs in light of current fuel mandates and for rationalizing the existing policies so as to direct a larger share of public resources to more promising options than corn ethanol, including cellulosic ethanol, biobutanol, and clean diesel fuel from organic waste. Other hurdles to biofuels deployment also need to be addressed, including deployment of critical supporting infrastructure. Finally, steps should be taken to ensure that policies aimed at reducing the nation's oil dependence do not end up promoting unsustainable fuel alternatives.

On technology innovation, the Commission reiterated the recommendations from the 2004 report that had not been enacted. It called again for a doubling of federal expenditures on energy technology research, development, and demonstration—and a tripling of international cooperation—with an emphasis on public-private partnerships and technologies that offer high leverage against multiple challenges. "In other words, let's aim at oil dependence and climate change at once with technologies that do both."

Holdren showed the effect that the Commission's recommendations would have on carbon dioxide emissions until 2030 (Figure 12.5). The original set of proposals would stabilize emissions at 8,000 million metric tons of carbon diox-

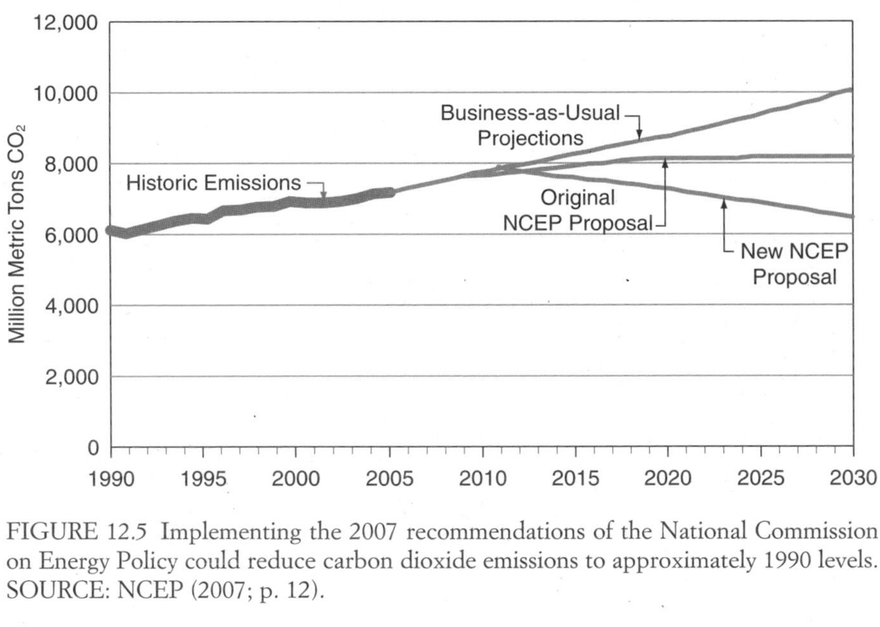

FIGURE 12.5 Implementing the 2007 recommendations of the National Commission on Energy Policy could reduce carbon dioxide emissions to approximately 1990 levels. SOURCE: NCEP (2007; p. 12).

ide, whereas the new recommendations would cause emissions to drop during the period to approximately 1990 levels. The largest share of reductions would come from the electric power sector, with reductions from the transportation sector the second largest.

The December 2007 energy legislation enacted more of the steps called for by the Commission. It raised CAFE standards to 35 miles per gallon for the combined fleet of cars and light trucks by model year 2020. It also modified the Renewable Energy Standard to start at 9 billion gallons in 2008, rising to 36 billion gallons by 2022, with 21 billion gallons of that to come from advanced biofuels. New Energy Efficiency Equipment Standards cover lighting, residential refrigerators, freezers, and commercial walk-in coolers and freezers.

The 2007 legislation expanded research, development, and demonstration for carbon capture and sequestration, directed the Department of Energy to engage the National Academies to review the program, directed the Department of Energy to work with the National Academies to develop interdisciplinary graduate degree programs in geological sequestration, established a university-based research and development grant program to study carbon capture and sequestration with different types of coal, and created a new efficiency and conservation block grant to be funded at $2 billion per year for 5 years.

Several provisions that were in the House bill were not included in the enacted law. These include an energy portfolio standard of 15 percent for renewables by 2020 and a 4-year extension of the tax credit for renewable electricity production.

Two oil and gas subsidies were repealed in order to pay for the implementation of the CAFE provisions. However, most of the oil and gas subsidies were left unchanged. As Holdren said, "Tom Friedman gave a talk I heard a few weeks ago in which he said, 'You couldn't make this up. They finally passed a big energy bill, and they left the oil and gas subsidies in and took the renewables subsidies out.'" The House bill called for repealing $22 billion a year in tax subsidies for oil and gas to pay for efficiency and renewables incentives and the implementation of CAFE, but the Senate version repealed only $1 billion of subsidies—enough to pay for the CAFE implementation but not enough for the efficiency and renewables incentives. The Senate's excuse was that President Bush had threatened to veto the energy bill because of these provisions, so the Senate engaged in "pre-emptive concession," according to Holdren.

The best way to gauge the actions of the federal government is to "follow the money," Holdren said. For example, in the proposed fiscal year 2009 budget, the request for the Federal Railroad Administration falls by one-third, and grants to Amtrak fall by about a half billion dollars. "Why is that important?" Holdren asked. "It's important because rail is the most energy-efficient way to transport people and freight, and we're taking away the support for that transport mode." The budget request also eliminated the $227 million Weatherization Assistance Program. Energy technology research, development, and demonstration would increase by 6 percent in real terms, bringing it 44 percent above its fiscal year 2006 low. But within that amount, energy-efficiency research and development would go down 2.5 percent, and renewable-energy research, development, and demonstration would fall 28 percent compared to the enacted 2008 budget. "So the news is not, by any means, all good."

Over the past 8 years, funding for efficiency has gone down, fossil fuel funding has gone up, and hydrogen research is up, although down substantially in the fiscal year 2009 request (Table 12.2). Funding for nuclear fission is up substantially, but "a lot of that is, unfortunately, in programs that look designed to push toward early commercial reprocessing, which the Energy Commission thinks is a terrible idea." Renewables have gone up and down but, in the fiscal year 2009 request, are down.

Several bills addressing energy issues were introduced in the 110th Congress. Most are broadly compatible with the long-term goal of reducing U.S. greenhouse gas emissions by 60 to 80 percent by 2050. "The key point is that I don't think we ought to be arguing about whether we need 40 or 50 or 80 percent in 2050," Holdren said. "The curve looks the same for the next few years in any case. What we need to do is get on it."

Holdren acknowledged that the recommendations of the Commission are "too timid." The 2007 report admitted as much, saying that even if all the recommendations were enacted, carbon dioxide emissions would not fall to the levels they need to reach. However, the Commission's recommendations were designed to "reflect our best judgment of what could actually get enacted in the U.S. Congress," Holdren said, although "that, too, can be accused of being a

TABLE 12.2 Funding for Energy Research, Development, and Demonstration at the Department of Energy (million $), 2001 to 2009

	2001	2006	2008	2009[a]
Efficiency[b]	548.5	382.5	426.8	416.8
Fossil	518.1	472.8	554.3	598.2
Hydrogen	26.0	131.8	173.1	117.6
Nuclear fission	61.9	236.2	387.5	554.3
Nuclear fusion	236.4	241.1	235.5	396.6
Renewables	335.7	222.8	599.0	452.4
T&D	—	125.3	83.4	69.7
Total	1,726.5	1,812.6	2,461.5	2,605.5

[a]Administration budget request for 2009.

[b]Funding has fallen for efficiency since 2001 and has risen in other categories.

SOURCE: John Holdren, Harvard University, based on data from Kelly S. Gallagher, ETIP Energy RD&D Database, Energy Technology Innovation Project, Belfer Center for Science and International Affairs, Harvard Kennedy School, February 14, 2008, available at http://belfercenter.ksg.harvard.edu/publication/3238.

pre-emptive concession." The Commission decided that the greatest need is to get started and that more would be done over time. "We have to work in the society we live in, with the Congress we have," said Holdren.

For success in stabilizing carbon dioxide concentrations in the atmosphere, the greenhouse gas emissions of the United States should be turned around by 2012. "That would be my first measure of success," Holdren said: "You have to bend the curve. Wherever we need to end up by 2030 or 2050, we need to start bending that curve. . . . The Commission's recommendations would be enough to achieve that. And, again, if we can make it decline even faster, that's great."

CAP AND TRADE VERSUS A CARBON TAX

The Commission's most controversial recommendation in both its 2004 and 2007 reports was that there should be a safety valve on the greenhouse gas permit price. The concept is that if a predetermined price is reached in the marketplace, the government would sell as many additional permits as are demanded at that price. In other words, if the safety valve is triggered, the cap-and-trade system turns into a carbon tax. Companies, consumers, and other energy users would have to pay that price for each additional ton of carbon they emit.

The attraction of the safety valve is that it finesses a longstanding and otherwise irresolvable conflict between optimists and pessimists over the availability of affordable ways to reduce energy consumption, Holdren said. Pessimists fear that there will not be enough affordable options to reduce emissions cheaply and that the permit price will therefore be high, causing major economic dislocations. Optimists believe that even modest incentives will cause many consumption-reducing options to materialize. If the safety valve level has been reasonably set, it will not be reached and reductions in energy use will occur. "We would not have gotten this bipartisan, multisectoral group to agree to a mandatory economy-wide set of restrictions on greenhouse gases emissions in the United States without a safety-valve provision," Holdren said. "And many of us believe that some provision of this sort will be required to get the bipartisan support of the Congress to vote such a thing into existence."

However, the safety valve has been sharply criticized by some in the environmental community. "Some of my good friends in the environmental community have called me names for signing off on this proposition," said Holdren. The main criticism is that the safety valve unduly weakens the program by sacrificing the assurance that the stated target will be reached. The safety valve also is criticized as compromising market principles with what amount to price controls.

Holdren suggested that the key issues are the level of the trigger price and its rate of escalation over time. "If you set the safety valve too low or the escalation rate too slow, you will encounter the safety valve fairly early and you will end up with a carbon tax at a rather low value. But, if you set the thing right, that need not be the case." Holdren said that he is a technology optimist who believes that many energy-saving options will materialize once a reasonable incentive is in place. If that turns out to be wrong, Congress will certainly revisit the issue on a regular basis and make adjustments.

Also, the imposition of a carbon tax would not be a disaster. "Most economists start out with the view that a carbon tax would be preferable to cap and trade, but we accept cap and trade as second best because Congress is never going to pass something with the 'T' word in it. So, if what you end up with is a carbon tax by another name, why is that so horrible?"

The most important point is to get started sooner rather than later with a mandatory, economy-wide program. This became particularly apparent at the conference in Bali in December 2007, when many developing countries made it clear that they are ready to follow the lead of the United States and other developed countries. "Their rhetoric has changed dramatically in the last few years," said Holdren. "A few years ago, it was this is a problem that the industrialized countries mainly caused, and you're going to have to mainly cure [it], and how much we're willing to do about it is going to depend on how much you'll pay us. Now the rhetoric is, we understand that this is a problem that is affecting us. We understand that the ultimate solution will require our participation, because we've looked at the numbers and we understand that developing countries will be bigger than the industrialized nations as emitters of greenhouse gases after 2015."

For reasons of historical responsibility, capacity, equity, and international law, the United States and other industrialized nations should be first, Holdren said. "I believe that if the United States embraces mandatory economy-wide restraints, China and India and Brazil and Indonesia and Mexico will only be a few years behind in embracing mandatory economy-wide restraints. . . . That means it would be far better for the United States to get going sooner with something that is maybe a little less than ideal than to wait 2, 3, or 5 more years for a better proposal when the Congress, after additional evidence accumulates on the harm from climate change, might be willing to vote through a measure without a safety valve."

Other speakers at the summit made a strong case for a carbon tax instead of a cap-and-trade system. Paul Portney pointed out that past analyses have concluded that higher gasoline prices create a powerful incentive for automobile companies to make more fuel-efficient cars and for Americans to use the 200 million vehicles already on the road in more fuel-efficient ways. "For that reason, I think the prospect of a carbon tax—which would of course affect petroleum and gasoline prices and also coal and natural gas—is an attractive idea." A carbon tax could go up gradually, predictably, and over a considerable period of time. It would curtail not only the use of liquid fuels but also the use of natural gas and coal to produce electricity, while also reducing industrial, commercial, and residential use of fossil fuels. It would stimulate the production of alternatives to fossil fuels, including nuclear power, solar power, wind power, and geothermal power. It would reduce the nation's reliance on imports of petroleum and natural gas. It also could establish a floor underneath the prices of gasoline and coal if prices should collapse in the future, which cannot be ruled out if the U.S. and global economy were to falter.

Another reason for favoring a carbon tax, according to Portney, is that it would generate revenues. Though the U.S. economy has had full or nearly full employment for an extended period, the federal government is still generating large deficits. With a large cohort of people about to enter retirement age, they will begin to collect Social Security and Medicare, which will further add to the deficit. "I would rather raise these revenues by taxing something that we're trying to discourage than by raising income taxes or taxes on capital," Portney said.

Auctioning off permits under a cap-and-trade system would also generate revenue, Portney said, so another option would be a cap-and-trade system in which significant numbers of permits were sold. "But I think that most of the permits in the cap-and-trade system would inevitably be given away because politicians cannot resist the temptation to give away goodies. And for that reason it seems to me that a carbon tax is to be preferred."

Ged Davis noted that in many countries policy leaders are still engaged in an open discussion about the relative merits of a cap-and-trade mechanism or carbon taxation. He also noted that when markets are wide and deep, they can work well, but with relatively young markets there can be significant and sudden shifts in price. Such price fluctuations can reduce confidence for making long-term investments, which is an argument in favor of carbon taxes. However, "taxes can come and go depending on government positions," Davis acknowledged. And, as Portney noted, "like Count Dracula from a silver cross or the first rays of daylight, politicians run from any mention of new taxes, especially during an election year."

Holdren said that the fluctuations in carbon price in Europe were largely due to having set targets too low initially. The system underestimated how many opportunities were going to be available to avoid carbon emissions at a relatively low price. Also, there were problems with the allocation scheme and how many of the permits were given away. "I think we have all learned from that experience," Holdren said. "One of the reasons that the Energy Commission changed the details of its recommendations—including the fraction of the permits to be auctioned and the size and level of the safety valve—was learning from the European experience."

Another argument is that there should be not only a ceiling but also a floor on the permit price, Holdren said. Cap-and-trade approaches do generate variability in prices, which creates a lack of signals for firms that are trying to decide how much to invest in alternatives. More thinking needs to be done on the best way to construct a system that avoids pitfalls and maximizes benefits. That's one of the reasons that the Energy Commission is still in existence, Holdren said. It is continuing to work on questions involving cap-and-trade systems, the management of nuclear energy, and an expanded research and development program.

A single set of rules or policies will not work well everywhere, said Davis. What is needed is a wide range of suitable options, some of which will appeal in some countries, if not all. At the same time, new policies need to prove themselves within individual countries before they are adopted elsewhere in the world, and this process can take time.

13

The Role of the Private Sector—The Case of Google

In the last few years that Dan Reicher worked at the Department of Energy, he oversaw a budget of about a billion dollars as assistant secretary for Energy Efficiency and Renewable Energy. About that time he read an article in the *San Francisco Chronicle* saying that the venture capital community had spent a billion dollars in 2000 on clean energy technology. "I said to myself, why didn't I know anything about this? . . . I decided at that point I should go out and figure out this other point of the triangle."

He took a job helping to run a renewable energy company called Northern Power that was backed by venture capital. When that company was sold, he helped raise money from the California Teachers Retirement System and a large Silicon Valley venture capital firm to invest in a number of ethanol projects, biodiesel projects, wood-fired power plants, and co-generation projects, all with the goal of transitioning away from fossil fuels. "Across those 6 years, I got a better sense about not only how critical capital is to this transition, but [about] the fundamental connection of capital to policy and policy to technology."

In 2007 Reicher began working at Google on the company's clean energy and climate change initiatives. "This truly is an amazing moment for looking at a sustainable energy future," Reicher said. "Higher oil and gas prices. The climate challenge indeed turning into a climate crisis. Consumer imperatives reflected in the high cost of oil and gas. Many things are coming together that make this such an extraordinary moment. That's what was reflected in the decision by the co-founders of Google, Larry Page and Sergey Brin, to give this [issue] emphasis inside the company."

Several other considerations factored into the company's decision to emphasize energy, according to Reicher. Google is a significant user of electricity at its data centers in the United States and throughout the world. Furthermore, when Google has tried to procure green electricity for its data centers, the company has found it to be either nonexistent or very expensive. Also, the company has been reluctant to buy carbon offsets, since it has not been convinced that "when all is said and done [offsets] are really driving the kind of change we need."

Google also employs thousands of engineers, many of whom have mechanical, electrical, or energy backgrounds and are eager to work on energy challenges. With government and university research underfunded, and even with large amounts of venture money flowing into clean energy technologies, not enough funding is finding its way to high-risk and high-potential technologies. Plus, said Reicher, not enough of that money is finding its way to large-scale commercialization of high-risk technologies—the "so-called Valley of Death commercialization projects."

A sustainable energy future requires three elements, according to Reicher. Technologies need to be developed. Smart policies need to be put in place. And there must be adequate capital to make the multitrillion-dollar transition to sustainable energy systems.

RENEWABLE ENERGY

Google has launched two main initiatives. The first is called Renewable Electricity Cheaper Than Coal, or RE < C. The initiative reflects the prominent place of coal in electricity generation, both in the United States and around the world. Replacing or mitigating the effects of coal burning is "a big, big challenge in terms of the global climate crisis," Reicher said. Coal-fired plants currently account for 500,000 megawatts of capacity in the United States, compared to a total installed wind capacity of 12,000 megawatts. Furthermore, the capacity of coal-fired plants is projected to grow dramatically, both in the United States and in other countries. For example, China installed almost 100,000 megawatts of conventional coal-fired power plants in 2006 alone.

There is considerable optimism about the prospects for renewable energy from the public and the investment community. But renewable sources of energy are not now economical compared with coal, Reicher observed. On the contrary, the price of electricity from renewable sources can be as much as four to five times the cost of electricity from coal. "Our observation was that there needs to be a fundamental change in the cost structure of renewables if we really expect them to be able to compete."

RE < C has the goal of making renewable electricity competitive with coal, and in a period of years rather than decades. In particular, the project is seeking to generate from renewable resources a gigawatt of electricity—enough to power the city of San Francisco—that is cheaper than the electricity from coal.

The initiative is focusing on solar thermal energy, photovoltaic energy, wind power, enhanced geothermal systems, and electricity transmission and storage. In addition, it is approaching the problem through the triangle of technology, policy, and finance, Reicher said. First, Google is hiring scientists and engineers who will work on key renewable energy challenges. It is funding external research and development in universities and other institutions. And it has a "reasonably significant amount of money" to invest in high-risk technology companies and in high-risk early-stage commercialization projects.

Google also has considered how the information tools available at Google and elsewhere such as Google Earth, Google Maps, and You Tube could influence the use of energy. For example, Google Earth could be used to capture data about renewable energy resources. Then transmission resources, policy incentives, and other useful information could be layered on top of the resources information. Such a resource could provide information to everyone from consumers to technologists to policymakers. As another example, a pilot project using Google Maps available on a mobile unit can show a person where public transportation is available, how the travel time compares with a car's, and the relative carbon emissions between the two. "There are lots of interesting things that you can do to give people options."

The "smartness" in a home, a vehicle, or an electric grid is in many respects a function of information, Reicher said. He encourages people "to think about where information and energy intersect. Ultimately, I think that's our greatest potential."

So far Google has made just a few modest investments, with more in the works. For example, it has invested $10 million in an advanced solar thermal company called E-Solar, which is using large mirror-based systems to turn solar energy into steam and then electricity. It also has invested $10 million in a pioneering high-altitude wind energy technology company. The strongest wind speeds are not at 50 or 100 meters above the ground but at 5,000 to 10,000 meters and higher. If a way could be found to tap those winds, wind power could be much greater and more consistent. "So we put some money into a company . . . that is looking at some very interesting technologies, admittedly very high-risk technologies, that might allow us to capture this resource." The company also is interested in cellulosic ethanol, where large-scale plants will be hard to finance because of their high risk.

"We're doing what other venture capital entities do, which is go out and canvas the landscape worldwide looking for the most compelling investment opportunities. The difference with us is that we don't have some of the constraints of the venture world." Google does not need to see a return on investment in 3 to 6 years or have exit strategies in place. "We can tolerate higher risks in our investments [and] lower returns."

In the policy arena, the company is looking at traditional tools, but it also is asking itself whether there are ways to be more creative. For example, the federal system of research and development tax credits is highly inconsistent. They exist for a year or two and then need to be reauthorized, which sometimes happens and sometimes does not. While Google supports extending the renewable energy production tax credits, it also is asking the question, "Is a tax code the right way to be incentivizing renewable energy in this country?" Reicher said. Other countries take different approaches. For example, many European countries work through fees and tariffs. "We are asking ourselves whether there is another way to get at this." Another policy innovation would be to merge efficiency standards and renewable energy goals into a national renewable electricity standard.

An important need is to link federal support of research and development with private sector activities. Today the private sector investment in renewables is substantial, and results can be achieved more quickly and with less duplication if public and private sector investments are integrated. "To be very honest with you, I don't think Silicon Valley knows how to talk to Washington, D.C., and vice versa, very well." Few venture capital firms have spent time learning about the federal research and development system to understand the potential for collaboration. Also, Wall Street needs to understand the federal research and development system "because that's where the big long-term equity is . . . to get the big, big projects built once they come out of the pilot-scale test."

Finally, a cap on carbon emissions is important, said Reicher, but it is not enough. More direct and immediate policy options are essential. Once legislation is adopted and regulations are written to implement the legislation, years can pass. "If the Clean Air Act Amendments are any indication, we could be looking at a decade, and we don't have a decade to wait in terms of the climate crisis."

HYBRID ELECTRIC VEHICLES

The other major initiative going on at Google is known as RechargeIt, which focuses on accelerating the commercialization of hybrid electric vehicles. A small fleet of after-market Priuses and Ford Escapes are being driven by Google employees, with data from their use of the vehicles being made available on a publicly accessible website. One of the largest solar photovoltaic systems in the United States, covering a large parking structure and many roofs of the company, supplies electricity for the vehicles. The vehicles are getting 70 miles per gallon of gasoline equivalent. That may not be as much as the 100 miles per gallon that people who have their own particular plug-in vehicle are able to get our of their vehicles, Reicher said. "But it's still significantly more than the 40 or 50 miles per gallon that the average Prius is getting today."

Many companies are interested in plug-in vehicle technology, Reicher said. When Google expressed an interest in investing in such technologies, it received almost 400 proposals from companies for equity investments. Google also is trying to create lines of communication that would be available to the major automakers. "The automobile industry is not used to talking to the electric utility industry," Reicher said. "They need to learn how to talk to each other because, ultimately, plug-in vehicles are going to be very much about a grid that can not only support but also enhance the value of these vehicles."

One major question is whether the electric grid is capable of supporting millions of vehicles. A study of the issue found that there is adequate capacity in California for approximately 4 million plug-in electric vehicles, if the vehicles are charged largely at night. A separate study by Jon Wellinghoff of the Federal Energy Regulatory Commission found that it would be economical not only for the grid to power plug-in hybrids but also for the vehicles to return power to the grid. For example, when Google launched its plug-in hybrids, it did an experiment with Pacific Gas and Electric. The company sent a signal to one of the cars over the Internet telling it to stop charging and start sending electricity back into the grid. "Fortunately, because all the press was there, it worked."

Admittedly, said Reicher, that is just an experiment, and lots of debate surrounds the idea. But if millions of vehicles were available that could provide energy to the grid, the cars could produce as much as $2,000 to $4,000 per vehicle, according to Wellinghoff's analysis. "Thus the notion of a cash-back hybrid," Reicher said. Such a system would be especially well suited for California, because the wind tends to blow later in the day and not during peak demand. "But imagine if you had millions of vehicles that could in fact be charged with this wind-generated electricity and then through a more intelligent grid sell that electricity back. It's a very exciting deployment."

If renewable electricity cheaper than coal could be combined with plug-in vehicles, this country could have much more confidence about its ability to deal with the climate crisis and energy security, Reicher said. The nation would burn less coal and oil and dramatically cut greenhouse gas emissions. "It's a world that Google is excited about. It's a world that Google is very committed to help create working with others."

14

Barriers and Potential

Senator Jeff Bingaman challenged the participants at the summit to identify which American President said the following words:

> I am inaugurating a program to marshal both government and private research with the goal of producing an unconventionally powered, virtually pollution-free automobile within five years.

It was President Nixon, in a February 10, 1970, special message to the Congress on environmental quality. In fact, as early as 1958, President Eisenhower proposed limits on the importation of foreign oil, James Schlesinger noted. These calls have led to a succession of federal programs addressing various aspects of energy policy (Figure 14.1). Yet they have not had the impact on energy supplies or uses that their proponents envisioned.

One reason for the limited success of these programs is that the United States does a good job of developing technologies but is less adept at commercializing those technologies, Bingaman said. For example, the Clinton-era Partnership for a New Generation of Vehicles spent about $300 million a year in research and development on such technologies as hybrid engines. But many of the resulting technologies were brought to market by Japanese rather than U.S. firms.

Successive administrations have an incentive to show that a new initiative is a significant improvement over the program of a predecessor, Bingaman said. In doing so, the progress made by a predecessor is sometimes abandoned. New programs also tend to be marked by excessively optimistic assumptions

Vehicle Technology

- Virtually pollution-free car (Nixon 1970)
- Reinventing the Car (Carter 1977-1980)
- Partnership for a New Generation of Vehicles (Clinton 1993-2000)
- FreedomCar (Bush 2003)

Coal Utilization

- Synthetic Fuels Corporation (1979-1985)
- Clean Coal Technology Program (1987)
- Clean Coal Power Initiative (2001)
- FutureGen (2003)

Nuclear Technology

- Clinch River Breeder Reactor (1970-1983)
- Advanced Liquid Metal Reactor Program (1989-1994)
- Global Nuclear Energy Partnership (2006)

Biofuels

- Alcohol fuels (Energy Security Act 1980)
- Oxygenated fuels (Clean Air Act Amendments 1990)
- Biofuels (EPAct 2005; EISA 2007)

FIGURE 14.1 Starts and stops in energy technology policy. NOTE: EPAct 2005, Energy Policy Act of 2005; EISA 2007, Energy Independence and Security Act of 2007. SOURCE: Senator Jeff Bingaman, U.S. Senate Energy and Natural Resources Committee.

about the costs and capabilities of particular technologies, while not enough consideration is given to the interplay of such programs with other policy areas. And such programs often demonstrate an underappreciation of the scale of the energy enterprise, especially given the difficulties of introducing new technologies on a large scale.

A president has to be thoroughly engaged in the energy problem and willing to put political capital behind it, said Schlesinger. Few presidents are that engaged in energy issues. The one possible exception was President Carter, who unsuccessfully proposed substantial taxation on the use of hydrocarbons, "and so far as I've been able to see," Schlesinger said, "most politicians have not wanted to emulate him."

The success of governmental policies depends on how well the political process works, Schlesinger observed. But the U.S. government is characterized by a separation of powers, which was designed to avoid a concentration of power that might be dangerous to individual liberties. One consequence of this arrangement is continuous disputation among the three branches of government. "Blame it on King George III," said Schlesinger.

Politicians also are loath to bring bad news to the public. Schlesinger quoted Russell Long, chair of the Energy Committee, to the effect that the first rule of a politician is to get elected and the second rule is to get re-elected.

"That does not lead to a great deal of courage," said Schlesinger. Politicians tend to engage in tokenism and symbolism. They "put forward something that seems to be moving toward the goal that people are concerned about, but it does very little since it involves very little sacrifice." For example, various governors have pledged that they will reduce greenhouse gases by 90 percent by the year 2080. "They will not be in office when that day comes around," observed Schlesinger.

Also, politicians tend not to think quantitatively, Schlesinger said. They do not consider the vast amounts of effort required to reduce the more than 7 billion barrels of oil that the United States uses each year or the amounts of greenhouse gases released. And bureaucracies change very slowly, Schlesinger observed. "How high is the Department of Homeland Security on the pecking order amongst government departments at the moment?" he asked. "Homeland security is a very important issue, but [DHS] doesn't have as much clout as some of the older departments."

Holdren pointed out that the status quo is supported by powerful and wealthy interests. For that reason, those interests tend to be preserved in the policy process. Breaking the hold of the status quo will require several major changes, he said. Partnerships that keep businesses viable can be an agent of change. For example, carbon capture and sequestration will be needed to keep the coal-powered generation business viable. Coal companies recognize that fact and now support approaches that include regulations on carbon dioxide so long as they also include support for technology development and incentives to deploy those technologies. Energy companies also have lots of experience in removing liquids and gases from the ground and know how to put liquids and gases back in place. "They see this as a big emerging business," said Holdren. Another example is the interest of oil companies in biofuels, and especially advanced biofuels, that use some of the capabilities those companies have in chemical engineering.

Public education also is critical, Holdren said. Al Gore's documentary *An Inconvenient Truth* focused public attention on climate issues. Similarly, articulate spokespeople will increasingly point out that there are jobs to be created, money to be made, and competitive advantage to be grasped by figuring out how to deliver goods and services that people want in energy-saving ways.

Finally, campaign finance reform is essential to loosen the grip of entrenched interests. "Without campaign finance reform, we're not going to get as much done as we need to, because there are going to be elements of the status quo that are extremely well-funded and are going to continue to be able to buy enough votes to avoid change," Holdren said.

The structure of the federal government is another impediment to change, according to Portney. "I don't think that the U.S. government is organized in a way that suggests that we take energy as seriously as I think we should or that the government itself thinks [energy] should be taken." The Department of

Energy's budget—about $24 billion—may look impressive, Portney said. But much of that funding is devoted to the National Nuclear Security Administration and the cleanup of chemical and nuclear contamination at former and current defense weapons plants, leaving a relatively small percentage for energy issues.

In other parts of the federal bureaucracy, the National Highway Traffic Safety Administration in the Transportation Department oversees CAFE standards. The Minerals Management Service in the Interior Department is responsible for about a quarter of the oil that is produced in the United States on the outer continental shelf and for about 15 percent of the natural gas produced in the United States. The Minerals Management Service is also responsible for permitting offshore wind turbines. The Federal Energy Regulatory Commission, an independent regulatory body, is responsible for siting hydroelectric facilities and other important facilities in the United States, regulates pricing of interstate and wholesale electricity, oversees interstate shipments of natural gas and petroleum through pipelines, and has a number of other important responsibilities. The Nuclear Regulatory Commission, which is another independent regulatory agency, has responsibility for permitting and inspecting new and established nuclear reactors in the United States. The Internal Revenue Service in the Department of the Treasury is in the process of writing regulations about who is qualified to receive subsidies for ethanol and other renewables production under the recent energy bill. And the Environmental Protection Agency writes national ambient air quality standards, national emissions standards for hazardous air pollutants, and new source performance standards for new industrial facilities, including petroleum refineries; rates vehicle emissions standards; promulgates fuel recipes for different parts of the country at different times of the year; and establishes effluent standards for ethanol and other biofuel plants in the United States. "This is clearly the agency that has the biggest impact on energy in the United States," said Portney.

"I want to be very clear in saying that I'm not criticizing any of these agencies," Portney said. "They do—and do well—exactly what it is that Congress has told them to do under the laws that Congress has passed that give them their mandate. I am saying that if we are to treat energy in the United States as seriously as it deserves to be treated, we have to do better organizationally than the mishmash of agencies that we have attending to this critical problem now. And this will require changes in legislation, not just executive orders."

REASONS FOR OPTIMISM

Why should the potential for progress be different today than it has been in the past? Several speakers cited possible reasons for optimism. First, technologies that could make a difference in energy production and consumption are closer to the marketplace than they have been previously. Hybrids have entered the market, and plug-in automobiles are on the horizon. Technology development is getting a big boost from high gas prices, which have "stirred a lot of entrepreneurial juices," Schlesinger said. And discussions of carbon taxes or other means of internalizing the costs of greenhouse gas emissions raise the prospect that policy changes could give a substantial boost to clean technologies.

Also, the public and politicians may be more engaged in the problem today than in the past. "The public has to be hit over the head by a two by four" to focus on energy problems, Schlesinger observed. While he doubted that the public was as engaged today as it was during the 1973 and 1979 oil crises, he and other speakers pointed to a growing level of interest in the issue in the United States and other countries—sparked in part by rapid recent increases in energy prices.

Also, some speakers observed that the public and policymakers are more aware of the need to move forward on many fronts simultaneously. "The primary solution to the challenges of energy security and environmental preservation is energy diversification," said Reuben Jeffrey. "Greater diversity of energy types, sources, and distribution networks can help improve the security and reliability of energy supplies, mitigate the economic consequences of high oil prices, and promote responsible environmental stewardship."

As Samuel Bodman described the situation, "The bottom line is this: We are seeing a convergence of forces that tells me that our nation is on a path to a cleaner, affordable, and more diverse energy future. The rigorous debate and analyses that the Academies are fostering—and to which all of you are lending your extensive expertise—will help ensure that we continue on the right pathway toward a more secure energy future."

CONTRIBUTIONS FROM SCIENTISTS AND ENGINEERS

To reduce energy intensity, the contributions of scientists and engineers will be essential, Bingaman said. Ray Orbach reminded summit participants of President Bush's words in his 2008 State of the Union address: "To keep America competitive into the future, you must trust in the skill of our scientists and engineers and empower them to pursue the breakthroughs of tomorrow."

Scientists and engineers also need to participate in the shaping of public policies, Bingaman observed. Congress is continually bombarded with information, much of which has a significant bias. Furthermore, Congress does not have the luxury of waiting to act until it has perfect information. Legislation has to accommodate uncertainty and then be monitored to track the effects of laws.

Along with both the crafting of legislation and oversight of implementation, strong and balanced technical input is critical. In that regard, reports from the National Academies and other organizations provide critical input into policymaking, Bingaman said. Reports from these organizations offer balanced and complete analyses of difficult questions, add technical realism to energy policy, and avoid simplistic views, which helps Congress figure out which technologies will have the greatest benefit at the least cost. Recent controversies over the energy-saving potential of biofuels or of changes in daylight saving time are good examples of the need for reliable information, Bingaman observed. "I would feel much more comfortable knowing that we were basing policy decisions about energy and climate on the best information available." Policymakers also can use the information in these reports to push for more aggressive action by the federal government.

COMMUNICATING WITH THE PUBLIC

Another essential need is to communicate the urgency of the situation to the public. "Not enough of us spend enough time talking to the public," said John Holdren. "I suggested in my AAAS presidential address last year that everybody in the science and technology community who cares about the future of the world should be tithing 10 percent of his or her time to interacting with the public in the policy process on these issues, and a lot of that is just giving talks. Wherever I go and give talks about these issues, the reaction is: 'I didn't know that. I had no idea how big this problem was. What are the opportunities for addressing it?' If all of us just got out there and talked to the public more and talked to policymakers more, we would get some of this done."

Dan Reicher emphasized that a focus on solutions is often more productive than a focus on problems. "The public has, to some extent, become numb about the problems," he said. People realize that the challenge is immense, but they do not realize how straightforward and cost-effective many of the solutions are. For example, many people can be motivated to take action, Reicher said, by providing them with information about the energy used in their homes and businesses and laying out a plan for no-cost, low-cost, and medium-cost initiatives.

However, it is also important to emphasize that the problems will not be easy to solve, several speakers said. "We always have to communicate that we can do it, but it's not going to be easy, . . . and it's going to cost," said Steven Specker. Also, all options need to be considered. "We need everything," said Specker. "Each country, each state, each community may decide to deploy or not deploy certain technologies, but as technologists, our message has to be that we need to be working on everything."

In an uncertain world, no single approach is guaranteed to work. Some currently promising technologies will not work, and others will arise that are not currently anticipated. "As many people have observed, there is no single magic bullet," said Harold Shapiro.

References

CCTP (U.S. Climate Change Technology Program). 2006. *Strategic Plan*. DOE/PI-005. U.S. Department of Energy, Washington, D.C. Available at http://www.climatetechnology.gov/.

Cheah, L., C. Evans, A. Bandivadekar, and J. Heywood. 2007. *Factor of Two: Halving the Fuel Consumption of New U.S. Automobiles by 2035*. Laboratory for Energy and the Environment, Massachusetts Institute of Technology, Cambridge, Mass. Available at http://lfee.mit.edu/public/factor_2_final_report.pdf.

Deutch, J., and E. Moniz. 2003. *The Future of Nuclear Power: An Interdisciplinary MIT Study*. MIT Press, Cambridge, Mass. Available at http://web.mit.edu/nuclearpower/.

Deutch, J., and E. Moniz. 2007. *The Future of Coal—Options for a Carbon-Constrained World*. MIT Press, Cambridge, Mass. Available at http://web.mit.edu/coal/.

EIA (Energy Information Administration). 2007. *International Energy Outlook 2007*. Office of Integrated Analysis and Forecasting, U.S. Department of Energy. Available at http://www.eia.doe.gov/oiaf/ieo/pdf/0484(2007).pdf.

EIA. 2008. *Annual Energy Outlook 2008*. Office of Integrated Analysis and Forecasting, U.S. Department of Energy. Available at http://www.eia.doe.gov/oiaf/aeo/pdf/0383(2008).pdf.

Goldemberg, J., S.T. Coelho, O.S. Lucon, and P.M. Nastari. 2003. Ethanol learning curve—The Brazilian experience. *Biomass and Bioenergy* 26(3):301-304.

IAC (InterAcademy Council). 2007. *Lighting the Way: Toward a Sustainable Energy Future*. Amsterdam. Available at http://www.interacademycouncil.net/?id=12161.

IEA (International Energy Agency). 2006. *World Energy Outlook 2006*. Organisation for Economic Co-operation and Development, IEA, Paris. Available at http://www.worldenergyoutlook.org/2006.asp.

IEA. 2007. *World Energy Outlook 2007*. Organisation for Economic Co-operation and Development, IEA, Paris. Available at http://www.iea.org/Textbase/npsum/WEO2007SUM.pdf.

INL (Idaho National Laboratory). 2006. *The Future of Geothermal Energy: Impact of Enhanced Geothermal Systems (EGS) on the United States in the 21st Century*. DOE/INL Report No. INL/EXT-06-11746. INL, Idaho Falls, Idaho.

IPCC (Intergovernmental Panel on Climate Change). 2000. *Special Report on Emissions Scenarios*. Available at http://www.ipcc.ch/pdf/special-reports/spm/sres-en.pdf.

IPCC. 2007. *Climate Change 2007: Synthesis Report. Contribution of Working Groups I, II and III to the Fourth Assessment Report of the Intergovernmental Panel on Climate Change.* Core Writing Team, R.K. Pachauri and A. Reisinger, eds. IPCC, Geneva, Switzerland.

Macedo, I.C., MR.L.V. Leal, and J.E.A.R.da Silva. 2004. *Assessment of Greenhouse Gas Emissions in the Production and Use of Fuel Ethanol in Brazil.* Secretariat of the Environment of the State of São Paulo, Brazil. Available at http://www.unica.com.br/i_pages/files/pdf_ingles.pdf.

McKinsey & Company. 2007. *Reducing U.S. Greenhouse Gas Emissions: How Much at What Cost?* Available at http://www.mckinsey.com/clientservice/ccsi/pdf/US_ghg_final_report.pdf.

NCEP (National Commission on Energy Policy). 2004. *Ending the Energy Stalemate: A Bipartisan Strategy to Meet America's Energy Challenges.* Available at http://www.energycommission.org/ht/d/sp/i/492/pid/492.

NCEP. 2007. *Energy Policy Recommendations to the President and 110th Congress.* Available at http://www.energycommission.org/ht/display/ContentDetails/i/1577/pid/493.

NPC (National Petroleum Council). 2007. *Hard Truths: Facing the Hard Truths About Energy.* Available at http://www.npchardtruthsreport.org/.

NRC (National Research Council). 2008a. *Review of the Research Program of the FreedomCAR and Fuel Partnership: Second Report.* The National Academies Press, Washington, D.C. Available at http://books.nap.edu/openbook.php?record_id=12113.

NRC. 2008b. *Transitions to Alternative Transportation Technologies—A Focus on Hydrogen.* The National Academies Press, Washington, D.C. Available at http://books.nap.edu/openbook.php?record_id=12222.

NRC and NAE (National Research Council and National Academy of Engineering). 2004. *The Hydrogen Economy: Opportunities, Costs, Barriers, and R&D Needs.* The National Academies Press, Washington, D.C. Available at http://www.nap.edu/openbook.php?isbn=0309091632.

Pasternak, A. 2000. *Global Energy Futures and Human Development: A Framework for Analysis.* Lawrence Livermore National Laboratory Report No. UCRL-ID-140773. October.

Raupach, M.R., G. Marland, P. Ciais, C. Le Quéré, J.G. Canadell, G. Klepper, and C.B. Field. 2007. Global and regional drivers of accelerating CO_2 emissions. *Proceedings of the National Academy of Sciences USA* 104(24):10288-10293.

UK DTI (United Kingdom Department of Trade and Industry). 2003. *Technology Status Review and Carbon Abatement Potential of Renewable Transport Fuels in the UK.* Report B/U2/00785/REP, 2003. Available at www.dti.gov.uk/renewables/publications/pdfs/b200785.pdf.

UNICA (Brazilian Sugarcane Industry Association). 2005. *Sugarcane's Energy: Twelve Studies on Brazilian Sugarcane Agribusiness and Its Sustainability.* I. de Carvalho Macedo, organizer. Uniao da Agroindustria Canavieira de São Paulo, São Paulo.

USDA (U.S. Department of Agriculture). 1995. *Estimating the Net Energy Balance of Corn Ethanol.* Agricultural Economic Report No. 721, H. Shapouri, J.A. Duffield, and M.S. Graboski, Economic Research Service, Office of Energy. July. Available at http://www.ers.usda.gov/publications/aer721/AER721.PDF.

U.S. Geological Survey. 2000. *U.S. Geological Survey World Petroleum Assessment 2000—Description and Results.* U.S. Geological Survey World Energy Assessment Team. U.S. Geological Survey Digital Data Series—DDS-60.

Victor, D.G., J.R. Schlesinger, and J. Deutch. 2006. *National Security Consequences of U.S. Oil Dependency.* Council on Foreign Relations, Washington, D.C. Available at http://www.cfr.org/content/publications/attachments/EnergyTFR.pdf.

Wigley, T.M.L., R. Richels, and J.A. Edmonds. 1996. Economic and environmental choices in the stabilization of atmospheric CO_2 concentrations. *Nature* 379:240-243.

Appendixes

A

America's Energy Future Project

In 2007, the National Academies initiated a major study titled "America's Energy Future: Technology Opportunities, Risks, and Tradeoffs," to inform the national debate about the nation's energy future. The America's Energy Future (AEF) project is planned to include two phases (Figure A.1). Phase I will produce a series of five reports. The first is a summary of discussions at the National Academies' March 2008 energy summit convened to provide input to the deliberations of the Committee on America's Energy Future. Three reports produced by separately constituted panels will also provide material for consideration by the full AEF study committee in its report on the current and future potential of existing and new energy supply and demand technologies, their associated impacts, and projected costs. Phase I of the AEF project will serve as the foundation for a Phase II portfolio of subsequent studies at the Academies and elsewhere focused on more strategic, tactical and policy issues, such as energy research and development priorities, strategic energy technology development, policy analysis, and many related subjects.

A key objective of the AEF project is to facilitate a productive national policy debate about the nation's energy future.

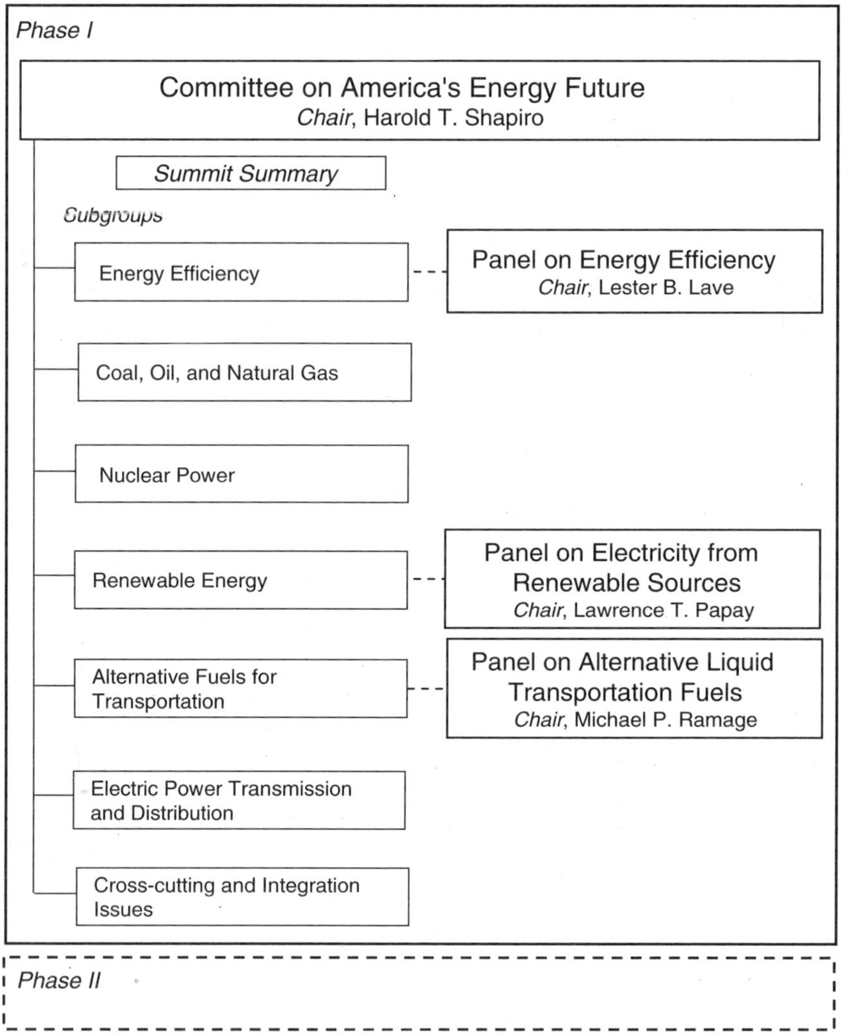

FIGURE A.1 America's Energy Future Project.

COMMITTEE ON AMERICA'S ENERGY FUTURE

HAROLD T. SHAPIRO (IOM), Princeton University, *Chair*
MARK S. WRIGHTON, Washington University, *Vice Chair*
JOHN F. AHEARNE (NAE), Sigma Xi, The Scientific Research Society
ALLEN J. BARD (NAS), University of Texas, Austin
JAN BEYEA, Consulting in the Public Interest
W.F. BRINKMAN (NAS), Princeton University
DOUGLAS M. CHAPIN (NAE), MPR Associates, Inc.
STEVEN CHU (NAS), E.O. Lawrence Berkeley National Laboratory
CHRISTINE A. EHLIG-ECONOMIDES (NAE), Texas A&M University, College Station
ROBERT W. FRI, Resources for the Future, Inc.
CHARLES H. GOODMAN, Southern Company Services, Inc. (retired)
JOHN B. HEYWOOD (NAE), Massachusetts Institute of Technology
LESTER B. LAVE (IOM), Carnegie Mellon University
JAMES J. MARKOWSKY (NAE), Independent Consultant
RICHARD A. MESERVE (NAE), Carnegie Institution for Science
WARREN F. MILLER, JR. (NAE), Texas A&M University-College Station
FRANKLIN M. ORR, JR. (NAE), Stanford University
LAWRENCE T. PAPAY (NAE), PQR, LLC
ARISTIDES A.N. PATRINOS, Synthetic Genomics, Inc.
MICHAEL P. RAMAGE (NAE), ExxonMobil Research and Engineering Company (retired)
MAXINE L. SAVITZ (NAE), Honeywell, Inc. (retired)
ROBERT H. SOCOLOW, Princeton University
JAMES L. SWEENEY, Stanford University
G. DAVID TILMAN (NAS), University of Minnesota, Minneapolis
C. MICHAEL WALTON (NAE), University of Texas, Austin

PANEL ON ENERGY EFFICIENCY

LESTER B. LAVE (IOM), Carnegie Mellon University, *Chair*
MAXINE L. SAVITZ (NAE), Honeywell, Inc. (retired), *Vice Chair*
R. STEPHEN BERRY (NAS), University of Chicago
MARILYN A. BROWN, Georgia Institute of Technology
LINDA R. COHEN, University of California, Irvine
MAGNUS G. CRAFORD (NAE), LumiLeds Lighting
PAUL A. DECOTIS, State of New York, Office of the Governor
JAMES DEGRAFFENREIDT, JR., WGL Holdings, Inc.
HOWARD GELLER, Southwest Energy Efficiency Project
DAVID B. GOLDSTEIN, Natural Resources Defense Council
ALEXANDER MACLACHLAN (NAE), E.I. du Pont de Nemours and Company (retired)

WILLIAM F. POWERS (NAE), Ford Motor Company (retired)
ARTHUR H. ROSENFELD, E.O. Lawrence Berkeley National Laboratory
DANIEL SPERLING, University of California, Davis

PANEL ON ELECTRICITY FROM RENEWABLE SOURCES

LAWRENCE T. PAPAY (NAE), PQR, LLC, *Chair*
ALLEN J. BARD (NAS), University of Texas at Austin, *Vice Chair*
RAKESH AGRAWAL (NAE), Purdue University
WILLIAM L. CHAMEIDES (NAS), Duke University
JANE H. DAVIDSON, University of Minnesota, Minneapolis
J. MICHAEL DAVIS, Pacific Northwest National Laboratory
KELLY FLETCHER, GE Global Research
CHARLES F. GAY, Applied Materials, Inc.
CHARLES GOODMAN, Southern Company Services, Inc. (retired)
SOSSINA M. HAILE, California Institute of Technology
NATHAN S. LEWIS, California Institute of Technology
KAREN L. PALMER, Resources for the Future, Inc.
JEFFREY M. PETERSON, New York State Energy Research and Development Authority
KARL RABAGO, AES Corporation
CARL J. WEINBERG, Pacific Gas and Electric Company (retired)
KURT E. YEAGER, Electric Power Research Institute, Inc. (retired)

PANEL ON ALTERNATIVE LIQUID TRANSPORTATION FUELS

MICHAEL P. RAMAGE (NAE), ExxonMobil Research and Engineering Company (retired), *Chair*
G. DAVID TILMAN (NAS), University of Minnesota, Minneapolis, *Vice Chair*
DAVID GRAY, Noblis
ROBERT D. HALL, CDG Management, Inc.
EDWARD A. HILER (NAE), Texas A&M University-College Station (retired)
W.S. WINSTON HO (NAE), Ohio State University
DOUGLAS R. KARLEN, Iowa State University
JAMES R. KATZER (NAE), Independent Consultant
MICHAEL R. LADISCH (NAE), Purdue University
JOHN A. MIRANOWSKI, Iowa State University
MICHAEL OPPENHEIMER, Princeton University
RONALD F. PROBSTEIN (NAS/NAE), Massachusetts Institute of Technology
HAROLD H. SCHOBERT, Pennsylvania State University

CHRISTOPHER R. SOMERVILLE (NAS), University of California, Berkeley
GREGORY STEPHANOPOULOS (NAE), Massachusetts Institute of Technology
JAMES L. SWEENEY, Stanford University

B

Speakers at the Summit

JEFF BINGAMAN is currently serving his fifth term in the U.S. Senate representing the state of New Mexico. He is chair of the Committee on Energy and Natural Resources and has been involved in all major U.S. energy legislation since his election to the Senate in 1982. Senator Bingaman also serves on the Senate Finance Committee, where he is chair of the Subcommittee on Energy, Natural Resources, and Infrastructure. He also serves on the Health, Education, Labor and Pensions Committee, and the Joint Economic Committee. Prior to his Senate career, Senator Bingaman served as New Mexico's Attorney General. After graduating from Harvard University, he earned a law degree at Stanford University and practiced law in New Mexico until his election as Attorney General.

SAMUEL W. BODMAN is the 11th U.S. Secretary of Energy. Prior to this, he served as an associate professor of chemical engineering at the Massachusetts Institute of Technology and began his work in the financial sector as technical director of the American Research and Development Corporation, a pioneer venture capital firm. From there, Secretary Bodman went to Fidelity Venture Associates, a division of Fidelity Investments where, in 1983, he was named president and COO and a director of the Fidelity Group of Mutual Funds. In 1987, he joined Cabot Corporation, where he served as chairman, CEO, and a director. He is a former director of MIT's School of Engineering Practice and a former member of the MIT Commission on Education. Secretary Bodman is a member of the National Academy of Engineering. He holds a B.S. degree in chemical engineering from Cornell University and a D.Sc. degree from Massachusetts Institute of Technology.

STEVEN CHU is director of the Lawrence Berkeley National Laboratory and professor of physics and molecular and cell biology at the University of California, Berkeley. He is a leader in U.S. and international energy science and technology communities and recently co-chaired the InterAcademy Council study, *Lighting the Way: Toward a Sustainable Energy Future*. Dr. Chu has numerous awards, including the 1997 Nobel Prize in physics, and is a member of the National Academy of Sciences. He received A.B. and B.S. degrees in mathematics and physics from the University of Rochester, a Ph.D. in physics from University of California, Berkeley, and 10 honorary degrees.

RALPH J. CICERONE is president of the National Academy of Sciences and chair of the National Research Council. He is an atmospheric scientist whose research in atmospheric chemistry and climate change has involved him in shaping science and environmental policy at the highest levels nationally and internationally. His research was recognized on the citation for the 1995 Nobel Prize in chemistry awarded to University of California, Irvine, colleague F. Sherwood Rowland. The Franklin Institute recognized his fundamental contributions to the understanding of greenhouse gases and ozone depletion by selecting Dr. Cicerone as the 1999 laureate for the Bower Award and Prize for Achievement in Science and recognition of his public policy leadership in protecting the global environment. Dr. Cicerone is a member of the National Academy of Sciences, the American Academy of Arts and Sciences, and the American Philosophical Society. He received his bachelor's degree in electrical engineering from the Massachusetts Institute of Technology and both his master's and doctoral degrees from the University of Illinois in electrical engineering.

JON CREYTS is a principal in the Chicago office of McKinsey & Company, Inc., which he joined in October 2000. He is the U.S. lead for the McKinsey Special Initiative on Climate Change and a co-leader of the global Capital Productivity Practice. Dr. Creyts has a concentrated knowledge of environmental management, capital productivity, plant operations, and fuel marketing and sourcing strategies and has served clients in the electric power, metals and mining, petroleum, travel and logistics, and retail sectors. Dr. Creyts received a B.S. degree from the University of Illinois and an M.S. and a Ph.D. in mechanical engineering from the University of California, Berkeley.

GED DAVIS is co-president of the Global Energy Assessment at IIASA in Laxenburg, Austria. Until March 2007 he was managing director of the World Economic Forum, responsible for global research, scenario projects, and the design of the annual Forum meeting at Davos. Before joining the Forum, Mr. Davis spent 30 years with Royal Dutch Shell. He was the vice president of global business environment for Shell International in London and head of Shell's scenario planning team. In this capacity he participated in a wide variety of global, regional, country, and industry scenario projects, many with a special focus on energy and environment. He has worked on many global projects for international institutions, including IPCC, IUCN, WBCSD, WEF, WEC, and UN agencies. He is currently advising a number of international institutions.

ROBERT W. FRI is a visiting scholar and senior fellow emeritus at Resources for the Future, where he served as president from 1986 to 1995. From 1996 to 2001 he served as director of the National Museum of Natural History at the Smithsonian Institution. Before joining the Smithsonian, Mr. Fri served in both the public and private sectors, specializing in energy and environmental issues. In 1971 he became the first deputy administrator of the Environmental Protection Agency. In 1975, President Ford appointed him as the deputy administrator of the Energy Research and Development Administration. He served as acting administrator of both agencies for extended periods. He received his B.A. degree in physics from Rice University and his M.B.A. degree (with distinction) from Harvard University. He is vice chair of the National Research Council's Board on Energy and Environmental Systems and recently chaired the National Research Council's Committee on Review of the DOE Nuclear Energy Research and Development Program.

KELLY SIMS GALLAGHER is director of the Energy Technology Innovation Policy research group at Harvard University's Belfer Center for Science and International Affairs and an adjunct lecturer in the Kennedy School of Government. She is an international member of the Task Force on Innovation for the China Council International Cooperation on Environment and Development. She recently published *China Shifts Gears: Automakers, Oil, Pollution, and Development*. Formerly, she was the science policy director of Ozone Action in Washington, D.C. She participated in more than a dozen rounds of international negotiations on global climate change and ozone depletion, and was an advisor to CNN in Kyoto and Buenos Aires for the climate negotiations. She was previously a Truman Scholar in the Office of Vice President Gore and worked in strategic planning at the international engineering and construction firm, Fluor Daniel. She has an A.B. degree in international affairs and environmental studies from Occidental College, and an M.A. in law and diplomacy and a Ph.D. in international affairs from the Fletcher School of Law and Diplomacy at Tufts University.

JOSE GOLDEMBERG is the secretary for the environment for the State of São Paulo, Brazil, and co-chair of the Global Energy Assessment Council of the International Institute for Applied Systems Analysis. He has served as the president of the Brazilian Association for the Advancement of Science, minister of state for education of the Federal Government of Brazil, and secretary for the environment of the State of São Paulo. He has authored many technical papers and books on nuclear physics, sustainable development, and energy. Dr. Goldemberg co-chaired the 2007 InterAcademy Council study, *Lighting the Way: Toward a Sustainable Energy Future*. In 2000, he was one of four recipients of the Volvo Environmental Prize presented in Sweden. A native of Brazil, Dr. Goldemberg earned his Ph.D. in physical science from the University of São Paulo.

JOHN P. HOLDREN is the Teresa and John Heinz Professor of Environmental Policy and director of the Program on Science, Technology, and Public Policy at the Kennedy School of Government, Harvard University, as well as president and director of the Woods Hole Research Center. From 1994 to 2001 he was a member of President Clinton's Committee of Advisors on Science and Technology and led major studies for the White House on U.S. energy research and development strategy, nuclear nonproliferation, and international cooperation on energy. Since 2002 he has been co-chair of the independent, bipartisan National Commission on Energy Policy and co-led the National Commission's project resulting in the major report, *Ending the Energy Stalemate.* He is also the coordinating lead author of the Scientific Expert Group on Climate Change and Sustainable Development. Dr. Holdren is a member of the National Academy of Sciences and the National Academy of Engineering. He received his B.S. and M.S. degrees in physics from the Massachusetts Institute of Technology and a Ph.D. from Stanford University.

REUBEN JEFFERY III is the under secretary, economic, energy and agricultural affairs at the U.S. Department of State, where he serves as the senior economic official. Dr. Jeffery advises the secretary of state on international economic policy. He leads the work of the State Department on issues ranging from trade, agriculture, and aviation to bilateral relations with America's economic partners. Dr. Jeffery received his B.A. degree in political science from Yale University in 1975 and J.D. and M.B.A. degrees from Stanford University in 1981.

AMORY LOVINS co-founded the Rocky Mountain Institute in 1982 and serves as its chief executive officer, research. An experimental physicist educated at Harvard and Oxford, Dr. Lovins rose to prominence during the oil crises of the 1970s when he challenged conventional supply-side dogma by urging that the United States instead follow a "soft energy path." His controversial recommendations were eventually accepted by the energy industry, and his book, *Soft Energy Paths: Toward a Durable Peace* (1977), went on to inspire a generation of decision makers. Dr. Lovins' work today focuses on transforming the car, real-estate, electricity, water, semiconductor, and several other manufacturing sectors toward advanced resource productivity. Since 1990, he has led the development of quintupled-efficiency, uncompromised, competitive automobiles and a profitable hydrogen transition strategy. Dr. Lovins was principal investigator for a major report released in 2004, *Winning the Oil End Game*.

ROBERT MARLAY is the deputy director of the U.S. Climate Change Technology Program and is a career member of the government's Senior Executive Service. He has more than 30 years of federal service and has been with the U.S. Department of Energy and its predecessor agencies since 1974. His contributions have focused primarily in the areas of national security, energy policy, science policy, and management of research and development programs. He serves concurrently as the department's director for the Office of Science and Technology Policy, Office of Policy, and International Affairs. He holds a B.S. degree in engineering from Duke University and a Ph.D. from the Massachusetts Institute of Technology. He is a licensed professional engineer in the District of Columbia. Dr. Marlay was a key contributor to the 2006 interagency report *U.S. Climate Change Technology Program Strategic Plan*.

RICHARD A. MESERVE is the president of the Carnegie Institution for Science. He served as chair of the U.S. Nuclear Regulatory Commission (USNRC) from 1999 to 2003, and he led the USNRC in responding to the 9/11 attacks. Dr. Meserve was elected to serve as a director of PG&E Corporation and Pacific Gas and Electric Company in 2006 and of Luminant Holding Company in 2008. He also serves on an advisory board to UniStar Nuclear Company, a joint venture of Constellation Energy and Électricité de France that seeks to market and operate nuclear power plants. Among other affiliations, he is a member of the National Academy of Engineering. Dr. Meserve received his undergraduate degree from Tufts University and his J.D. degree from Harvard University and a Ph.D. degree in applied physics from Stanford University.

ERNEST J. MONIZ is the Cecil and Ida Green Professor of Physics and Engineering Systems, director of the Energy Initiative, and director of the Laboratory for Energy and the Environment at the Massachusetts Institute of Technology, where he has served on the faculty since 1973. Dr. Moniz served as under secretary of the Department of Energy from 1997 until January 2001 and, from 1995 to 1997, as associate director for science in the Office of Science and Technology Policy in the Executive Office of the President. Dr. Moniz received a B.S. degree summa cum laude in physics from Boston College and a doctorate in theoretical physics from Stanford University. Dr. Moniz was a co-chair of the recent major interdisciplinary Massachusetts Institute of Technology studies that produced the reports *The Future of Nuclear Power* and *The Future of Coal*.

RODNEY NELSON is vice president of strategic marketing for Schlumberger Oilfield Services. Previously, Mr. Nelson was president of Schlumberger Data and Consulting Services where he directed operations of the industry's largest data processing and geotechnical consulting organization for clients worldwide. Mr. Nelson holds a bachelor's degree in engineering from the University of Wisconsin and an executive M.B.A. from the Massachusetts Institute of Technology's Sloan School of Management. Mr. Nelson was the Technology task group chair of the major effort completed in 2007 by the National Petroleum Council, entitled *Facing the Hard Truths About Energy*.

RAY L. ORBACH is the U.S. Department of Energy's first under secretary for science. Prior to his appointment as under secretary he was the Department's 14th director of the Office of Science. Dr. Orbach has received numerous honors as a scholar, including two Alfred P. Sloan Foundation fellowships; a National Science Foundation senior postdoctoral fellowship at Oxford University; a John Simon Guggenheim Memorial Foundation fellowship at Tel Aviv University; the Joliot Curie Professorship at the Ecole Superieure de Physique et Chimie Industrielle de la Ville de Paris; the Lorentz Professorship at the University of Leiden in the Netherlands; the 1991-1992 Andrew Lawson Memorial Lecturer at University of California, Riverside; and the 2004 Arnold O. Beckman Lecturer in Science and Innovation at the University of Illinois at Urbana-Champaign. Dr. Orbach received his B.S. degree in physics from the California Institute of Technology and his Ph.D. degree in physics from the University of California, Berkeley, and he was elected to Phi Beta Kappa.

KENNETH J. OSTROWSKI is a director in the Atlanta Office of McKinsey & Company with more than 23 years of consulting experience. Mr. Ostrowski is the leader of McKinsey's North America Electric Power and Natural Gas (EPNG) Practice and co-leads the Global EPNG Practice. Over the course of his career, he has served electric power, natural gas, and industrial clients in refining their strategic aspirations and direction, and in aligning the organizational, regulatory, and operational elements necessary to execute them. Before joining McKinsey, Mr. Ostrowski was an intern at the Congressional Budget Office and completed the 2-year Financial Management Program at General Electric Company. He received an M.B.A. in general management with honors from Harvard Business School and a bachelor's degree in finance, magna cum laude, from the University of Notre Dame.

PAUL R. PORTNEY is the dean and Halle Chair in Leadership of the Eller College of Management at the University of Arizona. From 1972 through June 2005, Dr. Portney was with Resources for the Future, where he headed two of its research divisions before becoming its vice president and then president and CEO. From 1979 to 1980, he also served as chief economist for the White House Council on Environmental Quality. Dr. Portney received his B.A. in economics from Alma College in Michigan and his Ph.D. in economics from

Northwestern University. Dr. Portney was chair of the National Research Council committee that produced the report *Automotive Fuel Economy: How Far Should We Go?*

MICHAEL P. RAMAGE is the retired executive vice president of ExxonMobil Research and Engineering Company. Previously he was executive vice president, chief technology officer, and director of Mobil Oil Corporation. He has broad experience in many aspects of the petroleum and chemical industries and is a member of Secretary of Energy Bodman's Hydrogen Technical Advisory Council. Dr. Ramage chaired the National Research Council study that produced the report *The Hydrogen Economy: Opportunities, Costs, Barriers, and R&D Needs* and is currently chairing the Committee on Resource Requirements for a Hydrogen Economy and the Panel on Alternative Liquid Transportation Fuel of the America's Energy Future effort. Dr. Ramage has B.S., M.S., Ph.D., and H.D.R. degrees in chemical engineering from Purdue University and is a member and former council member of the National Academy of Engineering.

DAN W. REICHER is the director of Climate Change and Energy Initiatives for Google. Prior to this Dr. Reicher served as president and co-founder of New Energy Capital Corp., a New England-based company that develops, invests in, owns, and operates renewable energy and distributed generation projects. He is currently a member of General Electric's Ecomagination Advisory Board, co-chair of the advisory board of the American Council on Renewable Energy, and a member of the board of the American Council for an Energy Efficient Economy. From 1997-2001, Dr. Reicher was assistant secretary of energy for energy efficiency and renewable energy at the U.S. Department of Energy. He holds a B.A. in biology from Dartmouth College and a J.D. from Stanford Law School. He also studied at Harvard's Kennedy School of Government.

MAXINE SAVITZ is vice president of the National Academy of Engineering, director of the Washington Advisory Group, and a member of the board of directors of the Charles Stark Draper Laboratories. She is a former deputy assistant secretary for conservation of the U.S. Department of Energy and received the Outstanding Service Medal from the DOE in 1981. Prior to her DOE service, she was program manager for research applied to national needs at the National Science Foundation. She recently retired from the position of general manager for technology partnerships at Honeywell after over 30 years of managing research, development, and implementation programs for the public and private sectors. She holds a Ph.D. in organic chemistry from the Massachusetts Institute of Technology and a B.A. degree in chemistry from Bryn Mawr College. Dr. Savitz is the vice chair of the Panel on Energy Efficiency Technologies of the America's Energy Future project.

JAMES R. SCHLESINGER is chair of the MITRE Corporation and a senior advisor for the investment banking firm, Lehman Brothers. He is a consultant to the U.S. Department of Defense, a member of the Defense Policy Board, a member of the Homeland Security Advisory Council, and director for Peabody Energy, KFx, Inc., and Sandia National Corporation. Dr. Schlesinger was the nation's first secretary of energy and also served as director of Central Intelligence, secretary of defense, and chair of the Atomic Energy Commission. Dr. Schlesinger received a B.A. degree summa cum laude from Harvard College, where he was elected to Phi Beta Kappa and was selected for the Frederick Sheldon Prize Fellowship. He received his M.A. and doctoral degrees in economics from Harvard University.

HAROLD T. SHAPIRO is president emeritus of Princeton University and a professor of economics and public affairs in the Woodrow Wilson School of Public and International Affairs. He served as president of Princeton from 1988 to 2001 and president of the University of Michigan from 1980 to 1988. Dr. Shapiro chaired the National Bioethics Advisory Committee under President Clinton and served on the President's Council of Advisors on Science and Technology under President Bush. He holds a bachelor's degree from McGill University and M.A. and Ph.D. degrees in economics from Princeton University as well as 14 honorary doctoral degrees. He is a member of the Institute of Medicine and the American Philosophical Society, and a fellow of the American Academy of Arts and Sciences and of the College of Physicians of Philadelphia. He is a member of the Board of Trustees of the Alfred P. Sloan Foundation, a member of the board of directors of Dow Chemical Company, and a member of the board of overseers of the Robert Wood Johnson Medical School. Dr. Shapiro is currently chairing the National Research Council study, "America's Energy Future: Technology Opportunities, Risks, and Tradeoffs."

STEVEN R. SPECKER is president and chief executive officer of the Electric Power Research Institute. Prior to joining EPRI, Dr. Specker served as president of Specker Consulting, LLC, where he provided operational and strategic planning services to technology companies serving the global electric power industry. In January 2003, he retired from General Electric as vice president of global marketing for GE Energy where he oversaw development of products and services for the generation, transmission, and distribution of electric power. He holds a B.S. degree in engineering science, an M.S. in nuclear engineering, and a doctorate in nuclear engineering from Iowa State University.

CHARLES M. VEST is president of the U.S. National Academy of Engineering and vice chair of the National Research Council. Prior to his election as NAE president, Dr. Vest served as president of the Massachusetts Institute of Technology from 1990 to 2004. During his tenure, the Massachusetts Institute of

Technology launched its OpenCourseWare initiative, co-founded the Alliance for Global Sustainability, enhanced the racial, gender, and cultural diversity of its students and faculty, established major new institutes in neuroscience and genomic medicine, and redeveloped much of its campus. Dr. Vest served on the President's Council of Advisors on Science and Technology during both the Clinton and Bush administrations, the Commission on the Intelligence Capabilities of the United States Regarding Weapons of Mass Destruction, the secretary of education's Commission on the Future of Higher Education, the secretary of state's Advisory Committee on Transformational Diplomacy, and the Rice-Chertoff Secure Borders and Open Doors Advisory Committee. He serves on the boards of several nonprofit organizations and foundations devoted to education, science, and technology. Dr. Vest has received honorary doctoral degrees from 10 universities and was awarded the 2006 National Medal of Technology by President Bush. He earned his B.S. degree in mechanical engineering from West Virginia University and an M.S. degree in engineering and a Ph.D. degree in mechanical engineering from the University of Michigan.

C

Summit Agenda

THURSDAY, MARCH 13, 2008

8:00 a.m. Welcome and Introduction
Ralph J. Cicerone, President, National Academy of Sciences

Current U.S. Energy Policy Context
Senator Jeff Bingaman, Chair, Committee on Energy and Natural Resources, U.S. Senate

8:30 Meeting Emerging Challenges to Global Energy Security
Reuben Jeffery III, Under Secretary for Economic, Energy and Agricultural Affairs, U.S. Department of State

9:00 The Geopolitical Context of America's Energy Future
James R. Schlesinger, Chair, The MITRE Corporation, and Senior Advisor, Lehman Brothers

9:30 Break

10:00 Summit Overview
Robert W. Fri, Senior Fellow Emeritus, Resources for the Future

---

NOTE: The National Academies Summit on America's Energy Future was held on March 13 and 14, 2008, in the auditorium of the National Academy of Sciences Building, 2100 C Street, N.W., Washington, D.C.

Session One
Moderator: Maxine Savitz

10:10 Introduction to Session One
Maxine Savitz, Vice President, National Academy of Engineering

10:20 Global Energy and Environment Projections: Next Steps
Ged Davis, Co-President, Global Energy Assessment, International Institute for Applied Systems Analysis, and former managing director, World Economic Forum

10:45 The Rise of China
Kelly Sims Gallagher, Director, Energy Technology Innovation Policy, Belfer Center, and Adjunct Lecturer at the Kennedy School, Harvard University

11:15 Facing the Hard Truths About Energy
Rodney Nelson, Vice President, Schlumberger, and Lead on Carbon Management, National Petroleum Council

11:45 Session One Question and Answer Forum

12:00 p.m. Lunch

Session Two
Moderator: Harold T. Shapiro

1:30 p.m. Introduction to Session Two
Harold T. Shapiro, President Emeritus, Princeton University, and Chair, National Research Council Committee on America's Energy Future

1:45 The Future of Coal and Nuclear Power
Ernest J. Moniz, Professor, Massachusetts Institute of Technology, and Co-Chair, MIT Interdisciplinary Studies on the Future of Nuclear Power and the Future of Coal Power

2:30 Biofuels: How Much, How Fast, and How Difficult?
José Goldemberg, Secretary for the Environment, State of São Paulo, Brazil, and Co-Chair, Global Energy Assessment Council, International Institute for Applied Systems Analysis

3:00 Break

3:30 Automotive Fuel Economy: How Far Should We Go?
Paul R. Portney, Dean, Eller College of Management, University of Arizona, and Chair, National Research Council Committee on Effectiveness and Impact of Corporate Average Fuel Economy Standards

4:00 Prospects of a Hydrogen Economy
Michael P. Ramage, Executive VP, ExxonMobil Research and Engineering Co. (retired), and Chair, National Research Council Committee on Alternatives and Strategies for Future Hydrogen Production and Use

4:30 Perspectives on America's Energy Future
Samuel W. Bodman, Secretary, U.S. Department of Energy
Moderator: Ralph J. Cicerone

5:15 Reception

FRIDAY, MARCH 14, 2008

Session Three

Moderator: Charles M. Vest

8:00 a.m. Introduction to Session Three
Charles M. Vest, President, National Academy of Engineering

8:15 Basic Research and America's Energy Future
Ray L. Orbach, Under Secretary for Science, U.S. Department of Energy

9:00 Ending the Energy Stalemate
John P. Holdren, Professor, Harvard University, and Co-Chair, National Commission on Energy Policy

9:45 Google's RechargeIT Program for Commercial Deployment of Plug-in Hybrid Vehicles
Dan W. Reicher, Director for Climate Change and Energy Initiatives, Google.org

10:15 Break

10:45	Electricity Innovation Pathways Steven R. Specker, President, Electric Power Research Institute
11:15	Session Three Question and Answer Forum
12:00 p.m.	Lunch

Session Four
Moderator: Richard A. Meserve

1:15	Introduction to Session Four Richard A. Meserve, President, Carnegie Institution for Science
1:30	Lighting the Way: Toward a Sustainable Energy Future Steven Chu, Director, Lawrence Berkeley National Laboratory, and Co-Chair, InterAcademy Council Study Panel on a Sustainable Energy Future
2:00	Reducing U.S. Greenhouse Gas Emissions: How Much at What Cost? Jon Creyts, Principal, McKinsey & Company Kenneth J. Ostrowski, Director, McKinsey & Company
2:30	Break
3:00	Winning the Oil End Game Amory Lovins, CEO, Rocky Mountain Institute, and Principal Investigator, *Winning the Oil End Game*
3:30	Climate Change Technologies Robert Marlay, Deputy Director, Climate Change Technology Program, U.S. Department of Energy
4:00	Session Four Question and Answer Forum
4:30	Closing Remarks and Adjourn Robert W. Fri, Senior Fellow Emeritus, Resources for the Future

D

Units of Measure and Equivalences

TABLE D.1 Prefixes for Units in the International System

Prefix	Symbol	Power	Value	Example
kilo	k	10^3	thousand	kilowatt (kW)
mega	M	10^6	million	megawatt (MW)
giga	G	10^9	billion	gigawatt (GW)
tera	T	10^{12}	trillion	terawatt (TW)
peta	P	10^{15}	quadrillion	petagram
exa	E	10^{18}	quintillion	exajoule

Units of Measure, Abbreviations, and Equivalences

Barrel (bbl) of oil = 0.136 tons of oil
British thermal unit (Btu) = 1,055 joules (J) = 2.93×10^4 kWh
CO_2 e = carbon dioxide equivalent
gallon (gal) = 3.785 liters
hectare (ha) = 10,000 square meters = 2.47 acres
kilowatt-hour (kWh) = 3.6×10^6 J
liter (l) = 0.2642 gal U.S.
metric ton (tonne) = 1,000 kg = 1.1023 short tons
million barrels of oil per day = 2.24 EJ per year
quad = quadrillion (10^{15}) Btu = 1.055 exajoules (EJ)